Praise for *Defend[...]*

"In this remarkable book, Nicolette Hahn N[...] [...] environmentalist—one who is willing to dig deeply, challenge orthodoxies, and get to the truth. You should read *Defending Beef* not only for the compelling case she makes for sustainable meat production, but also as an example of critical thinking at its finest."
> —Bo Burlingham, editor-at-large of *Inc.* magazine and author of *Small Giants and Finish Big: How Great Entrepreneurs Exit Their Companies on Top*

"I have traveled to every state in the U.S. during both summer and winter and have seen the land in extensive rural areas. There are huge land areas in this country that cannot be used for crops. The only way to grow food on these lands is by grazing animals. Grazing done properly will improve the land. *Defending Beef* shows clearly that beef cattle are an important part of sustainable agriculture."
> —Temple Grandin, author of *Animals Make Us Human* and professor of animal science, Colorado State University

"Anyone hesitating to eat beef due to environmental or nutritional concerns needs to learn the other side of the story. *Defending Beef* is both scientifically accurate and highly readable. Kudos to Nicolette Hahn Niman for successfully engaging in one of the biggest environmental tensions of our day."
> —Joel Salatin, farmer and author

"Creating healthful, delicious food in ecological balance is among humanity's greatest challenges. In this insightful book, Nicolette Hahn Niman shows why cattle on grass are an essential element. Every chef in America should read this book."
> —Alice Waters, executive chef, founder/owner, Chez Panisse, Berkeley, CA

"Anyone who doubts that beef can be part of a sustainable food system and healthy diet should read this book. *Defending Beef* proves beyond a shadow of a doubt that we can feel good about eating beef that's raised the right way."
> —Steve Ells, founder and CEO, Chipotle Mexican Grill

"Nicolette Hahn Niman just became beef's most articulate advocate. In *Defending Beef,* she pivots gracefully between the personal and the scientific, the impassioned and the evenhanded. It's a deeply compelling and delicious vision for the future of food."
> —Dan Barber, chef/co-owner, Blue Hill and Blue Hill at Stone Barns

"*Defending Beef* is a brave, clear-headed, and necessary addition to the discussion about sustainable food systems. Using hard data and solid scientific research, Nicolette Hahn Niman, a lawyer turned rancher, presents a convincing case that everything we thought we knew about the environmental and human health damage caused by beef is just plain wrong."
—Barry Estabrook, author, *Tomatoland: How Modern Industrial Agriculture Destroyed Our Most Alluring Fruit*

"The prosecution will never rest after the case presented here by this unusually well-armed defense lawyer. Exactly how much and in what ways cattle benefit our world—whether or not we eat beef—have never been more thoroughly explained. Cattle are lucky to have such a remarkable rancher gal come to their aid on our behalf."
—Betty Fussell, author, *Raising Steaks: The Life and Times of American Beef*

"Nicolette Hahn Niman's *Defending Beef* is as timely as it is necessary. With patience and passion she separates truth from fiction in the emotional debate about the role of beef in our lives and the effect of its production on our planet. Far from being a bogeyman of climate change and other environmental concerns, she argues, cattle, when properly managed, can play an important role in local food systems, land health, and carbon sequestration. The key is treating cattle as an ally, not an enemy, and exploring opportunities instead of simply pointing fingers. In this exploration, *Defending Beef* leads the way!"
—Courtney White, founder and creative director, Quivira Coalition, and author, *Grass, Soil, Hope*

"In our collective confusion and desperation about the environment, many zero in on cattle as an unlikely culprit for everything from water pollution to climate change. In *Defending Beef,* author, rancher, and environmental lawyer Nicolette Hahn Niman takes a nuanced look at the impact of livestock on land, water, the atmosphere, and human health. With clarity and eloquence, she puts research in context and shows that the raising of cattle can be destructive or restorative, depending on how the animals are managed. Cattle—and common sense—have found their champion."
—Judith D. Schwartz, author, *Cows Save the Planet*

"Issues related to the long-term health effects of red meat, saturated fat, sugar, and grains are complex and I see the jury as still out on many of them. While waiting for the science to be resolved, Hahn Niman's book is well worth reading for its forceful defense of the role of ruminant animals in sustainable food systems."
—Marion Nestle, professor of nutrition, food studies, and public health at New York University and author of *What to Eat*

"I hope this book, which is more about the future of humanity, will be read by every citizen—not just those who feel the need to defend their meat-eating preferences. Biologist, environmental lawyer, and mother Nicolette Hahn Niman has provided a balanced report on the effects of cattle production on our environment, health, and climate change. Openly accepting the damage done by modern-day cattle production—on the land and in factory feedlots—she effectively argues that cattle themselves are not the problem; it is the way they are being managed that is endangering our health, environment, and economy. We can do something about that, and we must for the sake of our children and grandchildren. Key to our success will be an informed citizenry—for whom this book will be an invaluable tool."

—Allan Savory, founder and president,
the Savory Institute

Defending Beef is an important book. Nicolette Hahn Niman had me at the chapter 'All Food Is Grass,' where she unpacks the complex clash of views over animal rights, ecology, and the legacy of human impact upon bioregions. The more I read, the more I came to value the passion and insight of someone who (like me) does not herself consume meat but recognizes that it rests at the center of what's troubling with our food system and how we might set it right.

"At Slow Food, we believe that better, less meat should become a rallying cry for a shift in our relationship to animals and each other. Scale, biodiversity, and rural economies get ample attention in this comprehensive yet easy-to-digest manifesto. If we ever hope to challenge the prevailing culture of confinement that defines the industrial meat system today, then we need to make this book required reading for butchers, bakers, and policymakers."

—Richard McCarthy, executive director,
Slow Food USA

"*Defending Beef* clearly and unequivocally connects the dots for us on how vitally important raising pastured beef is to humanity. From increasing the glomalin in soil that helps create healthy grass, to sequestering carbon, battling desertification, enhancing the water supply, mitigating climate change, and promoting biodiversity, Nicolette Hahn Niman carefully draws a constellation for understanding just how our food production systems affect people, culture, and our ecosystem—for good or ill. The case is airtight and the jury is in: Cattle on pasture are an integral part of the solution."

—Mary R. Cleaver, owner/executive chef, The Cleaver
Company and The Green Table, New York City

"A breakthrough book that reclaims our relationship with farm animals and nutritious food. Comprehensive and insightful, *Defending Beef* delivers a compelling description of a food system that works with nature and wildlife, supports humane animal husbandry, and builds strong local economies. With a keen mind and passionate love of life, Nicolette Hahn Niman provides an insightful solution to feeding our growing world population and shows us a way of life that is both beautiful and sustaining."

—Judy Wicks, founder of White Dog Café and the
Business Alliance for Local Living Economies
and author of *Good Morning, Beautiful Business*

"It is so important that we free our minds of conventional beef wisdom and open up to the solution set that uses nature's wisdom as well as the smart agricultural practices of the future. In *Defending Beef,* Nicolette Hahn Niman gives us an exacting and compelling defense of land management that solves for environmental resiliency, human health, climate change mitigation, and prosperity. How could we not listen?"

—Kat Taylor, CEO and cofounder, Beneficial State Bank; cofounder
and director, Tomkat Ranch Educational Foundation

"As a chef, I am concerned with not just the flavor of my ingredients, but also their ecological, economic, social, biodiversity, and health implications. In *Defending Beef,* Nicolette Hahn Niman delves deeply into the many impacts of beef production. Through both scholarly research and her own personal journey, she shows how, again and again, the 'conventional wisdom' has missed the mark, while making an extremely convincing case for well-raised cattle having a necessary place in our global agriculture system and on our plates. Simply put, this book doesn't just make me a better chef, but also a better person."

—Michael Leviton, chef/owner, Lumiere;
chair, Chefs Collaborative

"Nicolette Hahn Niman, a lawyer, vegetarian, and cattle rancher, serves up a well-argued defense of an American icon: the hamburger. Passionate and persuasive, Hahn Niman delivers a tough-minded critique of industrial animal operations along with an eloquent case on behalf of pasture-raised beef. The good news? It's safe to eat steak again—so long as you know where it comes from."

—Marc Gunther, editor-at-large,
Guardian Sustainable Business US

DEFENDING BEEF

THE CASE FOR SUSTAINABLE
MEAT PRODUCTION

NICOLETTE HAHN NIMAN

Chelsea Green Publishing
White River Junction, Vermont

Project Manager: Bill Bokermann
Project Editor: Benjamin Watson
Copy Editor: Laura Jorstad
Proofreader: Helen Walden
Indexer: Lee Lawton
Designer: Melissa Jacobson

Printed in The United States of America.
First printing October, 2014.
10 9 8 7 6 5 4 3 2   14 15 16 17

Chelsea Green Publishing is committed to preserving
ancient forests and natural resources. We elected to print
this title on 100-percent postconsumer recycled paper,
processed chlorine-free. As a result, for this printing, we
have saved:

38 Trees (40' tall and 6-8" diameter)
17 Million BTUs of Total Energy
3,235 Pounds of Greenhouse Gases
17,546 Gallons of Wastewater
1,175 Pounds of Solid Waste

Chelsea Green Publishing made this paper choice because
we and our printer, Thomson-Shore, Inc., are members
of the Green Press Initiative, a nonprofit program dedicated to supporting authors, publishers, and suppliers
in their efforts to reduce their use of fiber obtained
from endangered forests. For more information, visit:
www.greenpressinitiative.org.

Environmental impact estimates were made using the Environmental Defense Paper Calculator.
For more information visit: www.papercalculator.org.

Our Commitment to Green Publishing

Chelsea Green sees publishing as a tool for cultural change and ecological stewardship.
We strive to align our book manufacturing practices with our editorial mission and to
reduce the impact of our business enterprise in the environment. We print our books and
catalogs on chlorine-free recycled paper, using vegetable-based inks whenever possible.
This book may cost slightly more because it was printed on paper that contains recycled
fiber, and we hope you'll agree that it's worth it. Chelsea Green is a member of the Green
Press Initiative (www.greenpressinitiative.org), a nonprofit coalition of publishers, man-
ufacturers, and authors working to protect the world's endangered forests and conserve
natural resources. *Defending Beef* was printed on paper supplied by Thomson-Shore that
contains 100% postconsumer recycled fiber.

Library of Congress Cataloging-in-Publication Data
Niman, Nicolette Hahn.
  Defending beef : the case for sustainable meat production / Nicolette
Hahn Niman.
      pages cm
  Includes bibliographical references and index.
  ISBN 978-1-60358-536-1 (pbk.) — ISBN 978-1-60358-537-8 (ebook) 1.
Beef cattle—United States. 2. Ranching—Environmental aspects—United
States. 3. Diet—United States. I. Title.

  SF207.N56 2014
  338.1'76213—dc23

                              2014029308

Chelsea Green Publishing
85 North Main Street, Suite 120
White River Junction, VT 05001
(802) 295-6300
www.chelseagreen.com

For Miles and Nicholas

May you always appreciate cattle
for the food
and the life
they've provided you.

CONTENTS

PREFACE

IN 1970, AMERICA'S GRASSROOTS environmental movement was burgeoning as 20 million people poured into the streets to mark the first Earth Day on April 22. Cattle raising was dragged as a villain into the public square along with our nation's worst industrial polluters. Beef was increasingly regarded as an ecosystem destroyer and a primary cause of starvation around the globe; it was becoming part of the zeitgeist to believe that no genuine environmentalist or humanitarian would eat beef (at least not in a well-lit public place). Kicked off by *Diet for a Small Planet*, three decades of influential books like *Diet for a New America* and *Beyond Beef* then succeeded in making it nearly incontrovertible environmental gospel that beef is public enemy number one.

As a freshman biology major in the mid-1980s, I drank the Kool-Aid. I quit eating meat and enthusiastically embraced the attitude that no beef was good beef. Then I promptly filed the matter away; no more thought on the topic seemed necessary.

That logic fractured soon after I was hired as an environmental lawyer by Robert F. Kennedy Jr. He charged me with starting a national campaign on meat industry pollution. Initially, my assignment neatly reinforced my long-held negative views about meat and how it was produced. But the more farms I saw, the more studies I read, the more experts I interviewed, the less satisfied I was with my understanding of meat's connections to the environment. I began to recognize my views were simplistic—black-and-white and formed mostly from the bullet points in vegetarian and environmental pamphlets I'd encountered in college.

Fortunately, by the time a handsome cattle rancher proposed marriage to me two years later, my understanding of the role farm animals can and do play in food systems and natural ecosystems was far more nuanced. And I had the good sense to accept him.

Working in the meadows and valleys of our ranch alongside my husband for the past decade has given me an entirely new understanding of how natural environments work. I've lived among not only cattle but also domesticated and wild turkeys, deer, coyotes, newts, bobcats, ravens, hawks, egrets, gophers, and countless other animal and plant species. I've learned how humans can interact with ecosystems as landscapes that produce food while at the same time supporting—even enhancing—wild plant and animal populations that belong here.

A singularly negative view of ranching and beef persists among many environmentalists and among those who oppose raising animals for food. It's leaching into mainstream conversations as global warming concerns have infused the issue with new life. As a longtime environmentalist and vegetarian, I am intimately familiar with the criticisms. Rarely, though, have I encountered credible responses, least of all from the beef industry itself. Yet now—as someone who remains ecologically minded while raising cattle myself—I feel compelled to respond, honestly and passionately. This is my answer. This is my defense of beef.

INTRODUCTION

WE'VE ALL HEARD THE NARRATIVE so often—the one about how red meat, and beef in particular, is killing us—that many of us have come to accept it as incontrovertible truth. It's so common that it's become common knowledge. The story goes like this:

> Americans once raised cattle, pigs, and sheep on small, mixed farms scattered around the country and sprinkled with handfuls of livestock. Animal numbers were low and, correspondingly, Americans ate little red meat. We were thin. Hypertension, stroke, and heart disease rates were low. Environmental damage from farming was minimal. Over the course of the 20th century, however, everything changed for the worse. Livestock herds ballooned. Cattle began overgrazing the western half of the United States. Red meat and animal fat became abundant, cheap, and ubiquitous. Americans gorged themselves on hamburgers, butter, and ice cream. The result: soil erosion, water and air pollution, and skyrocketing rates of obesity and chronic diet-related diseases.

There's just one problem with this narrative: it's not true.

Yes, parts are correct. But facts that rarely make it into mainstream discussions and media coverage diametrically oppose key elements of this narrative. As this book will make clear, aspects of the United States' environmental condition have worsened, and chronic diet-related diseases have become more widespread and severe, but these problems cannot reasonably be connected with cattle or attributed directly to butter or beef. Why?

Because there are fewer cattle on the land today than there were a century ago. And because today we are eating less red meat in general, and less beef in particular, than at any time in recent history. We are also consuming less butter, far less whole milk, and much less saturated animal fat. No swelling bovine herds. No ever-heftier helpings of red meat and animal fat. The simplistic narrative completely collapses.

If you are skeptical, I won't blame you. What I've just said probably runs counter to what you've heard from innumerable sources for many years. But I come armed with data, and plenty of it, all from official government sources. While my overall premises—that cattle are good for the environment and that beef and animal fat are healthy food—are, admittedly, controversial in this day and age, the basic agricultural and demographic facts are not in dispute.

Here is the most pertinent bit of information to keep in mind. In the second half of this book, I will detail how American dietary choices have changed. I will show that we eat less beef and less animal fat now than we did 100 years ago, while our rates of grain and sugar consumption have skyrocketed. I will present facts strongly supporting the conclusion that our sugar and flour consumption rather than red meat and animal fat are to blame for the sharp rise in chronic diseases.

The popular narrative is also far off-base concerning the numbers of animals on the land. In reality, decreasing per-capita consumption has run parallel with a downward trend in cattle in U.S. inventory. The total quantity of red meat and dairy produced has increased in tandem with our rising population, and some is exported. But the amount of beef and dairy the United States exports is actually quite small. Only about 7 percent and 2 percent, respectively, go to foreign markets. So cattle raised for exported meat and milk products barely affect the math.

The bottom line is that increased production levels in both the beef and dairy sector have not been accompanied by expanding herd sizes. On the meat side, this is because animals are slaughtered at much younger ages. At the dawn of the 20th century, a typical beef steer going to slaughter was four or five years old.[1] Today, to lower costs, and enabled by growth hormones, that steer is killed at less than two years old, typically around 14 months.[2] Dairy cows, too, go to slaughter at much younger ages (often at just three years of age). This also affects beef supply because, now as always, a large portion of U.S. beef comes from dairy cattle.

The rise in milk production, however, is owing to an entirely different issue. As I detailed in my book *Righteous Porkchop*, selective breeding of dairy cows for greater milk output (read: large udders) has vastly increased per-animal production.[3] At the beginning of the 20th century, U.S. average per-cow milk production was 2,902 pounds annually. Today it is 19,951 pounds (about 2,347 gallons) per year. Of course, this is often touted as a huge victory for humanity. But the scale of the increase (nearly sevenfold) suggests that selective breeding has been pushed to an extreme. (Indeed, many of today's mature dairy cows even have trouble walking.) The net effect of this change has been a substantial diminishment in the size of the U.S. dairy herd over the past century.

These factors, combined with the virtual disappearance of animal draft power from agriculture, means there are a lot fewer large animals on the farms of the United States now than there were a century ago.

For those of you who may still find this difficult to believe, here are the specific numbers. Work animals on farms (horses, mules, and oxen) have gone from 22 million in 1900 to 3 million in 2002. In those same years, while beef cattle numbers went up, the increase was relatively modest, going from 44 million to

59 million. Sheep numbers plummeted, going from a high of 46 million in 1940 to 5 million today. Pig numbers have gone up, but only slightly: In 1920, there were 60 million pigs; in 2010, 65 million. The total herd of dairy cows shrank dramatically, from 32 million down to 9 million, over the past century. All told, that means that while early-20th-century farms and ranches had 76 million cattle and 164 million large animals, today they have 68 million cattle and 141 million large animals. In total, that's a 14 percent reduction in large farm animals.

From an environmental standpoint, two factors are significant: How many animals are in inventory at any given moment, and by what methods are they being raised? These factors will largely determine the sector's environmental footprint. From a human health and dietary standpoint, the central question is: What are we eating, and in what form? On the land, cattle are less present today than they were for most of the 20th century. On our plates, there is less beef and less bovine fat.

It is vitally important to understand these facts before continuing. If we recognize that cattle numbers and beef consumption are both down, it casts serious doubt on the all-too-common narrative that holds cattle responsible for our current environmental problems and beef for today's public health crisis. I would not expect these facts, on their own, to dissolve the concerns of beef's critics. But it is my hope that clarifying these facts at the outset will assist you in considering this book with an open mind.

While I strongly disagree with the conclusion that cattle and beef are responsible for America's environmental and health problems, aspects of the narrative I mentioned at the outset are correct. Serious environmental degradation has been caused by agriculture, including the cattle and beef sectors. Huge changes have indeed taken place in the way this country farms and eats. These issues will be explored in the pages of this book.

As I'll discuss in more detail, America's food and farm policies subsidize bad farming practices while providing little support for those that benefit the environment. They subsidize the production of unhealthy foods we are already overeating and on which we are getting sick. We continue on the path of overproduction. All of which results in underemployment, cheap food, overeating, waste, and pollution.

This book is at once a defense of cattle and beef, and an indictment of many aspects of the modern diet and modern agriculture. Change in both arenas is urgently needed. The United States is the world's top beef producer.[4] We can, and should, lead the world in forging a more environmentally sound way of raising cattle.

Whether you are among the critics or the defenders of beef, if you come along on this journey, you will find things to agree with and disagree with along the way. Whatever your perspective at the start, I hope by the end you will see things in a new and different light.

CATTLE

ENVIRONMENT AND CULTURE

CHAPTER 1

THE CLIMATE CHANGE CASE AGAINST CATTLE:

Sorting Fact from Fiction

FOR DECADES, THE PRIMARY OBJECTION to beef heard in environmental circles was "overgrazing." Cattle, it was said, had devastated vast stretches of the American West and much of the globe with their munching mouths and trampling hooves. Damaged waterways, eroded and denuded landscapes, reduced wildlife populations, and, most worryingly, deserts everywhere spreading like wildfires were the results. Deforestation, like that occurring in the Amazonian rain forest, was often tossed into the mix.

Two baseline notions always undergirded the charges against cattle—things so obvious to everyone they needed no proof. First, trampling and grazing inherently cause environmental injury: The more grazing and trampling, the more damage. Second, it was assumed that the more cattle are in an area, the graver the damage. A bit later, I will show how these beliefs all turn out to be incorrect. For the moment, I just want to point out these assumptions.

As a nature-minded college student, and later as an environmental lawyer, I fully accepted these ideas as I heard peers and colleagues decry dismal cattle-catalyzed chains of ecological destruction. The accusations are compelling, especially when accompanied by vivid images of barren lands out west and Brazilian forests being cut and burned, apparently to make way for cattle. I recall such photos playing a decisive role in my own choice to give up beef—the first meat I swore off—during my freshman year of college.

In the coming pages, I will address all the common criticisms of beef. But I want to begin with the burning issue cited most often in today's environmental literature: climate change. Intense media focus has sensationalized the matter by repeatedly brandishing sexy (yet silly) questions like: *Which is worse for the climate— eating a hamburger or driving a Hummer?* Usually such articles neatly conclude that the hamburger is more offensive and close by suggesting that swearing off beef is better than purchasing a Prius.

British journalist (and former editor at *The Ecologist*) Simon Fairlie dates the shift in rhetoric to 2007, since which time he notes that climate change "has become the main argument against carnivory." I've made the same observation on this side of the Atlantic.

Furthermore, like Fairlie, I consider a single document largely responsible. In late 2006, a division of the United Nations, the Food and Agriculture Organization (FAO), released a report titled *Livestock's Long Shadow* that blamed meat for 18 percent of greenhouse gas emissions, with cattle taking the lion's share of the blame. UN agencies release scores of reports every year; most garner scant public notice. Yet something about this report proved irresistible to the media—particularly the press release's headline, which cleverly compared emissions from cars with those from livestock.[1] By the following year, everyone from animal rights and environmental advocacy groups to the *New York Times* editorial board was treating the report's 18 percent figure as the gospel truth.

The earth has undoubtedly been subjected to a great deal of improper cattle grazing (far more than most ranchers like to acknowledge). And yes, cattle are indeed connected with climate change. But—as I will show—not in the ways people tend to assume. Regrettably, the recent flurry of attention to the cattle–climate connection has shed little light on the issue while generating a great deal of heat. In fact, these issues are very poorly understood by the general public, by environmentalists and animal activists, and even within the cattle and beef industries themselves.

FAO's *Livestock's Long Shadow* report only deepened the misperceptions. After it came out, I found myself increasingly frustrated with the widespread, inappropriate use of its 18 percent figure. Ultimately this prompted me to write an op-ed titled "The Carnivore's Dilemma," which *The New York Times* published in October 2009.[2] Nothing about livestock is *inherently* damaging to the environment, I argued. The problem lies instead with today's methods of raising them.

"The Carnivore's Dilemma" points out that when it comes to causes of climate change, the devil truly is in the details. To make sense of livestock's role in climate change, we must begin by considering individually the three main greenhouse gases: carbon dioxide (CO_2), methane (CH_4), and nitrous oxide (N_2O). For each, the issues are distinct.

Carbon dioxide makes up the majority of agriculture-related greenhouse gas emissions. In American farming, which is highly mechanized and automated, most carbon dioxide emissions come from the burning of fuel to operate vehicles and equipment. World agricultural carbon dioxide emissions, on the other hand, are mostly from a totally different source. Global emissions result primarily from the clear-cutting of woods, which is done for both crop growing and livestock grazing. During the 1990s, tropical deforestation in Brazil, India, Indonesia, Sudan, and other developing countries caused 15 to 35 percent (depending on the year) of annual global fossil fuel emissions, a portion of which is related to livestock.

In Brazil, the country with the most deforestation connected to livestock, swaths of tropical forest have been cut largely to create soybean fields. In Brazil's Mato Grosso state, soybeans are grown on nearly 70 percent of newly cleared areas. Over half of the nation's soy harvest is controlled by a handful of international agribusiness companies, which ship it worldwide for animal feed and other food products. Where forests are cut down to make way for soy fields, carbon dioxide is generated first from burning the trees and, later, from plowing, planting, harvesting, and transportation of the crops grown there.

However, no one eating beef need be part of this scenario. Beef cattle throughout the world are mostly kept on grass. The international soy market supplies industrial animal feeding operations in Europe, China, and, to a very small extent, the United States. Farmers and ranchers raising their cattle entirely on grass buy no soy, and cannot be blamed for these emissions. It is both inaccurate and unfair to impute such emissions to grass-raised cattle or those fed entirely with locally grown feeds.

Yet that is precisely what the FAO report did. It simply lumped all entities together and pinned deforestation emissions on an aggregated meat industry. The result—equally blaming farmers and ranchers the world over for the effects of deforestation taking place in very specific regions—is absurd.

Ironically, people who deliberately avoid beef may find it harder to avoid having their purchasing dollars funneled toward such deforestation. According to the Organic Consumers Association, soybeans from Brazil are found (unlabeled) in tofu and soy milk sold in American supermarkets.

Industrialized, mechanized facilities that confine animals around the clock—mostly dairy, pork, and poultry operations—do cause carbon dioxide emissions. These buildings have automated systems for feeding, lighting, sewage flushing, ventilation, heating, and cooling, all of which generate CO_2. These facilities, as well as beef cattle feedlots, must also continually provide feed for their animals, the growing, harvesting, drying, and transporting of which generate additional CO_2 emissions.

By contrast, pastoral livestock keepers create almost no carbon dioxide emissions. They are not mechanized; nor do they grow or buy feed. For all cattle, but especially those kept on grass, the CO_2 issue is negligible. When you take deforestation out of the equation, there is little connection between beef and carbon dioxide emissions.

Next, let's consider methane, agriculture's second-largest greenhouse gas. For starters, it should be noted that while methane emissions from beef and dairy cattle have been repeatedly singled out for criticism, other parts of the food system also generate substantial methane. Wetland rice fields, for example, caused as much as 29 percent of the world's total human-generated methane in the late 20th century.

In animal farming, liquefied manure storage lagoons are a major source of methane. Industrial pork, egg, and dairy facilities often add water to manure (for ease of transport) and hold millions of gallons in storage beneath confinement buildings and in ponds. This, too, is a problem uniquely associated with large-scale, industrialized agriculture. Neither smaller-scale nor grass-based operations use manure lagoons. "Before the 1970s, methane emissions from manure were minimal because the majority of livestock farms in the U.S. were small operations where animals deposited manure in pastures and corrals," U.S. Environmental Protection Agency documents state.[3] EPA research shows that methane emissions from livestock farming skyrocketed with the rapid rise of factory farms, as liquefied manure systems became the norm. Cattle ranches, and even beef feedlots, are not responsible for these emissions, as they do not use manure lagoons.

Beef's critics often point out that cattle are prime culprits in enteric methane production, and there is some truth to this. Viewed in a larger context, however, enteric emissions from cattle, which result from their unique, ruminant digestive system, should be regarded as a single side of a double-edged sword. Allow me to elaborate.

If a human ate a handful of grass, it would simply pass through her single-chambered stomach, providing no nourishment. A cow, however, has a symbiotic relationship with a complex of microorganisms—bacteria, protozoa, and fungi—residing in her four-chambered stomach. The digestive process involves fermentation in a chamber called a *rumen*, which is why cattle are called *ruminants*. The bovine gut flora are nothing short of miraculous: Tiny organisms aid digestion, enabling cattle to survive entirely on cellulosic plants like grass, which transforms the globe's vast grasslands into an invaluable piece of the human food system.

As a by-product of this unique digestive process, methane is burped (mostly), breathed out (somewhat), and farted (a little) by cattle. If you've ever stood near a cow when she masticates, you cannot help but observe this process. Methane results from the biological decomposition of the vegetation (some of which would occur, it should be noted, whether or not cattle were consuming it). Whenever methane from cattle is discussed, this duality should be kept in mind.

It's also important to remember that enormous herds of ruminant mammals have populated the globe for eons. In North America, for millennia prior to European colonization, grazing herds blanketed the continent, including millions of caribou and deer, an estimated 10 million elk, and somewhere between 30 and 75 million American bison. "The moving multitude . . . darkened the whole plains," Lewis and Clark wrote of bison in 1806. The total number of large ruminants was undoubtedly far in excess of the 40 million mature breeding beef cows and dairy cows, and even greater than the total 89.9 million cattle in the United States today.[4]

Moreover, although the digestive processes of cattle do generate methane, the causes of these emissions are now fairly well understood, and there are reasonable ways to curtail them. Methane production is aggravated when livestock eat poor-quality forages, throwing their digestive systems out of balance. (For this very reason, FAO's most recent climate report notes, cattle in the United States, which tend to have access to better forages, generate notably lower emissions than those in most other parts of the globe.[5]) Livestock nutrition experts have demonstrated that by making minor improvements in animal diets (such as by providing nutrient-laden salt licks), they can cut enteric methane by half.[6] Other practices, like adding certain proteins to ruminant diets, can reduce methane production per unit of milk or meat by a factor of six, according to research at Australia's University of New England.[7]

Myriad other practices have been found to reduce methane emissions from cattle. Research at the University of Louisiana has demonstrated that enteric methane emissions can be notably cut when cattle are regularly rotated onto fresh pastures. Other research at the world's universities has shown that breeding and feeding can lower enteric methane emissions, and that novel techniques like adding small weights to the rumen can reduce methane emissions by nearly a third. Surprisingly, research has even shown that the presence of dung beetles in meadows can significantly reduce the methane emitted from cattle on grass.[8] Much remains to be learned. But to me, these diverse and promising lines of research clearly suggest that bovine methane emissions are not an intractable problem.

Author Judith Schwartz has also pointed out that there is little correlation between methane levels and the number of ruminants in the world, a fact that raises doubt about their actual contribution to atmospheric methane. She cites a joint 2008 report from the FAO and International Atomic Energy Agency showing that since 1999 atmospheric methane concentrations have been stable, while at the same time the global population of ruminants grew at a rapid rate.[9]

Finally, agriculture also causes nitrous oxide emissions—but, as with carbon dioxide, the connection is weak. Nitrous oxide makes up around 5 percent of the total U.S. greenhouse gases.[10] More than three-quarters of the nitrous oxide emissions in agriculture result from use of manmade fertilizers. The compound is produced in the soil by two microbial processes: nitrification (ammonia oxidation) and denitrification (nitrate reduction). Research from the University of California–Berkeley in 2012, using a nitrogen isotope, proved for the first time that manmade fertilizers stimulate microbes in the soil to convert nitrogen to nitrous oxide at a faster-than-normal rate.[11]

Such research is damning to conventional, chemical-based agriculture (not just of livestock), but it is not a strike against beef. It is entirely possible to raise cattle (our own ranch is living proof) using few or no agricultural chemicals, and more and more people are doing it. Cattle that were neither fed crops grown with commercial fertilizer nor grazed on pastures where manmade fertilizers were applied have no connection with this chain of events.

In fact, studies like the Berkeley research actually provide further support for the importance of cattle and other livestock in the farming system. Animals offer the best ways to improve soils without chemical fertilizers. These methods include grazing animals on grass planted as part of crop rotations, grazing on crop residues, and applying animal manures to croplands.[12]

Now let's look more closely at the FAO report's 18 percent figure for meat's alleged contribution to global warming, which has some serious credibility problems.

For starters, this figure was always an outlier. As Simon Fairlie points out in his thoughtful and meticulously researched book, *Meat: A Benign Extravagance*, the 18 percent number far exceeded most other estimates from reputable scientific organizations. Among those were the Nobel Prize–winning Intergovernmental Panel on Climate Change (IPCC), which said at the same time that the *whole* of agriculture caused between 10 and 12 percent of global greenhouse gases.[13] Likewise, here in the United States, the Environmental Protection Agency has calculated that U.S. agriculture causes a total of 8 percent of U.S. global warming emissions.[14] It bears restating that this is for *all* U.S. agriculture, not just agriculture related to meat production. (By comparison, the EPA reports that U.S. transportation accounts for 28 percent of emissions.) Clearly, the FAO figure never reflected a scientific consensus, and it had limited application to animal farming here in the United States.

Additionally, unlike other sorts of climate change quantifications, such as total carbon entering the atmosphere, FAO's 18 percent figure cannot be regarded as objective, scientific data. Why? Because *Livestock's Long Shadow*, like many other reports, was crafted to support a particular policy agenda. In this case, the recommendations strongly suggest it was devised to strengthen the following argument: The global demand for meat is rising; confined pig and poultry operations have a lower climate change impact than cattle; thus the world food supply should move away from grazing animals and toward industrial poultry and pork.[15] (Remember, this is the report's perspective, not my own.) Even before its release, the report's lead author, Henning Steinfeld, a German agricultural economist, was on record making precisely this argument.[16]

I've personally met Steinfeld twice: once when he visited our California ranch, and once on a panel at a livestock conference in Bonn, Germany. From our direct conversations, as well from listening to him present his views at the conference, it was clear Steinfeld favors just such an approach. He believes that to fulfill a growing appetite for meat, the world should increase the number of large, confined pork and poultry operations while shrinking the herds of grazing animals inhabiting the world's grasslands.

I'm not suggesting Steinfeld is anything less than competent and sincere; he strikes me as both. But authors do have their perspectives. And understanding an author's point of view helps explain why certain things are included in a calculation while others are omitted. Such decisions are hotly debated in emerging fields like climate change. Henning Steinfeld views grazing animals as problematic, poultry and pork confinement operations less so. For reasons that will become clear in this book, I strongly disagree.

More recently, *Livestock's Long Shadow*'s 18 percent figure has fallen out of favor within the United Nations, even at FAO. In September 2013, FAO released a report it characterized as an "update" to *Livestock's Long Shadow*. In it, FAO revised the 18 percent figure down to 14 percent, a 22 percent reduction. Similarly, in November 2013, FAO's sister agency, the United Nations Environment Program, released a report called *The Emissions Gap Report 2013* which stated that for *all* agriculture the figure was just 11 percent.[17]

The 18 percent figure is dead and should never be cited again.

In addition to FAO, other attempts have been made to quantify livestock's global warming contribution, generally with even less credibility. Among them were two authors (neither scientists) who asserted in a 2009 Worldwatch publication that livestock were responsible for a whopping 51 percent of all global warming gases.[18] As one would expect, this bold pronouncement attracted considerable media attention.

However, the outrageous claim quickly crumpled under scrutiny, as it turned out the number was simply hatched by the authors. They acknowledged in interviews that the 51 percent figure was the result of neither new research nor even anecdotal information. Rather, the authors explained, they had read an article by a physicist who suggested including livestock respiration in greenhouse gas calculations. Then the authors simply took the figures from *Livestock's Long Shadow* and recalculated them, adding in a (large) number for animal respiration. Mind you, the physicist's article was also not reporting results of any climate change research; it was simply a two-page essay in which he espoused an idea. Hmm.

Notably, FAO's newest study (which concludes that livestock's role in greenhouse gases totals 14 percent rather than 18) *explicitly excluded* respiration. Hopefully it puts to rest the breathy idea once and for all. This must have shocked the authors of the Worldwatch piece, who had stated: "We heard recently that FAO have been sparked by our work to do their own recalculation . . . They have a lot of money and a lot of people; a lot of good mathematicians, and they have the world's best database on all these things, and so they're going to recalculate their own work and I'm sure that their 18 percent will move towards our 51 percent, or even exceed it."[19] Um, no.

As one would expect, the figures in *Livestock's Long Shadow* have also been dissected and criticized by those who feel they far overstate livestock emissions. In my estimation, the most important of these objections is the one I discussed earlier, that nearly half of the 18 percent number was due to carbon releases from deforestation in the developing world, particularly Brazil, India, Indonesia, and Sudan. The idea of including such emissions was novel, and it's the main reason FAO's number was so much higher than any other previous calculation. Yet, as I've suggested, including deforestation emissions raises several red flags.

In addition to the problems already highlighted, there's this one: A figure for numbers of acres cleared cannot be the basis for a credible annual figure of greenhouse gas emissions. As trees are felled and burned, carbon that was stored in trees is released to the air as CO_2. This is a onetime event. Once an area has been deforested it cannot be deforested again the next year, let alone the next. The CO_2 emitted from deforestation happens once, and that is all. To treat this as an annual, recurring figure is patently misleading. The fallacy in taking this approach is perfectly illustrated by FAO's own updated figure (14 percent), which was, according to the new report's authors, considerably lower primarily because Brazil has been taking serious measures to rein in its deforestation problem (although the 14 percent figure, too, includes the deforestation emissions in its total).

Moreover, as noted earlier, cutting down forests in the developing world, as troubling as it may be, has almost no connection to American beef consumption. True, in theory, soy from deforested areas could be imported as livestock feed for American cattle. In reality, however, the amount is infinitesimal. Less than one-third of 1 percent of all soybeans used in the United States are imported from *all* other countries. Meanwhile, three-quarters of Brazil's soybeans go to China, with nearly all of the rest going to the EU.[20] If any amount of soy at all goes from deforested Brazilian fields to American beef feedlots, it is statistically zero.

By the same token, only a minor percentage of all the beef and veal Americans consume (16 percent, according to USDA researchers[21]) is imported. And over 80 percent of that 16 percent comes from Canada, Australia, and New Zealand. In other words, the most that could possibly be imported from deforested developing countries is 3 percent. Thus, the environmental pamphlet depicting your hamburger as the chain saw razing the Amazonian rain forests is simply a fiction: Deforested lands in the developing world are neither the source of feed for U.S. cattle nor the source of cattle for American beef.

Meanwhile, in the United States trees are not cut down to make way for cattle, and they never were. Historically, forests in this part of the world were cleared for crop agriculture, timber, and railroads, not for grazing. Vast stretches of the United States—particularly in the Southeast, Midwest, and Far West—were grasslands when Europeans arrived. There is active debate over the contribution of Native Americans to the creation and maintenance of these grasslands. Regardless of any human role, land available for grazing was in place when Europeans began populating North America with their cattle in the 16th and 17th centuries. These existing grassy areas served as the primary grazing areas for both wild herbivores and domesticated livestock.[22] When humans cleared land of trees or actively prevented foresting, it was done for reasons other than grazing cattle.

I want to flesh out a bit the issue of Native Americans and the land because it's important to the broader questions of what North American lands looked like and how they functioned before cattle. A broad consensus now exists that Native Americans actively engaged in land management, including extensive burning, seed sowing, and horticulture. In his book on managed vegetation burning, University of California–Berkeley forestry professor Harold Biswell noted that this is borne out by firsthand observations of various explorers and naturalists.[23] Reasons for burning included the following: improving forage for wild grazing animals and hunting conditions; enhancing visibility for self-defense; facilitating travel; catching other animals, including grasshoppers, lizards, and snakes; improving seed production; clearing brush and undergrowth; and reducing fuels and fire hazards near settlements.[24]

Biswell additionally noted that, in California, burning was also used to ensure a good supply of acorns. Food sources were plentiful, and Indians found little need for agriculture. Acorns were their principal plant food, and oak trees were abundant over nearly all of California. "Oak tree preservation was perhaps one of the reasons for the Indians' use of fire," Biswell argues. "Not only do oaks show high resistance to surface fires, but their reproduction is favored by fire. The Indians probably knew this."[25]

In *1491*, Charles Mann's remarkable book about the Americas before the arrival of Columbus, the author writes that humans throughout the American continents made regular use of controlled burns. "[F]rom the Atlantic to the Pacific, from Hudson's Bay to the Rio Grande, the Haudenosaunee and almost every other Indian group shaped their environment, at least in part, by fire." In addition to the reasons previously mentioned, Mann argues that Native Americans engaged in "constant burning of undergrowth" that resulted in increased numbers of herbivores and the predators that ate them. "Indians retooled ecosystems to encourage elk, deer, and bear," Mann writes. He also points to historical evidence that Native Americans even deliberately vastly expanded the grazing range of the American bison to include areas from New York to Georgia.[26]

In his book *Dirt: The Erosion of Civilizations*, earth and space sciences professor David Montgomery also highlights the importance of active land management by early peoples around the globe. "Long before the last glacial advance, people around the world burned forest patches to maintain forage for game or to favor edible plants. Shaping their world to suit their needs, our hunting and gathering ancestors were not passive inhabitants of the landscape."[27]

When large numbers of Europeans came, they cleared American forests for lumber and for railroads, but mostly to grow crops. Livestock's use of the land was always secondary. On some farms, livestock grazed in rotations with crops or on land that had originally been cleared for crops. Montgomery documents widespread poor land stewardship as Europeans landed in the East and gradually migrated west. Land was cleared and plowed while crops were continuously planted (on southern plantations these were often large fields of cotton), without properly returning nutrients to the soils through cover crops, rotational grazing, or animal manures.

Such poorly managed fields became depleted of nutrients and severely eroded, eventually rendering them unsuitable for crop cultivation.[28] Livestock grazing followed as the default use for such damaged ground, giving the false impression that the damage from such poorly managed crop farming was caused by grazing.

Government records show that at the beginning of European settlement, about 46 percent of land that would become the United States was forested. By around 1900, it had been reduced to 34 percent. However, since that time some areas have reforested, making the overall number quite stable. Today 33 percent of U.S. land is forested.[29] In recent history, urban development, not agriculture, has been the primary threat to forested areas.

In any event, in recent decades the United States has not been losing forests; it has been gaining them. Total forested acres in the United States actually *increased* by 26 percent from 1990 to 2005.[30] Many U.S. cattle do graze in forests, and lately there is reason to believe that American ranchers, along with those in other parts of the world, will learn to create highly functioning ecosystems that involve both cattle herds and forests. With proper management, new research suggests, cattle and trees are not only compatible land uses, they are mutually beneficial. (The presence of cattle on the land also often enables landowners to keep the land from being cleared and converted to strip malls.)

This is now being demonstrated in a big way in Colombia, where farmers and ranchers are being encouraged to plant trees in pastures used for cattle grazing. "Cattle raised using silvopastoral techniques can digest the forage more easily and reduce their methane emissions by 20 percent," according to researcher Michael Peters of the International Center for Tropical Agriculture in Cali, Colombia.[31]

This is an important reversal. In recent history, agriculture departments encouraged American ranchers and farmers to remove oaks and other trees to increase cattle production.[32]

Systems combining cattle and trees have now been shown to be more profitable for ranchers and to enhance carbon sequestration in both trees and in the soils.

Such research serves as yet another reminder of how problematic it is to generalize about beef's contribution to global warming. Everything is highly site-specific. Keep in mind that nearly half of the FAO's original calculation for meat-related greenhouse gases was from deforestation. Yet I've shown that American beef has virtually no connection to deforestation emissions. Actually, telling an American to quit eating hamburgers because of deforestation is nonsensical.

Focusing narrowly on emissions from ruminants while failing to fully consider the larger role they play in ecosystems and agriculture is myopic. In response to FAO's 2006 report, the well-respected nonprofit UK Soil Association issued a comprehensive report on capturing and keeping carbon in soils. In it, the association notes that a recent European Commission report had concluded that European grasslands have the potential to sequester large amounts of carbon on an ongoing basis. In the UK, the sequestration of carbon in agricultural grasslands was calculated at around 670 kilograms of carbon per hectare per year. That amount "would offset all the methane emissions of [UK] beef cattle and about half those of dairy cattle."

In specific reference to Henning Steinfeld's preferred policy direction, the report states: "Advocates of a shift from red meat to grain-fed white meat to reduce methane emissions could therefore find that this has the perverse effect of exchanging methane emissions for carbon emissions from soils and the destruction of tropical habitats (to produce soya feed), as well as having a far reaching impact on our countryside, wildlife and animal welfare."[33]

This leads to my final critique of FAO's calculations, which also relates to the need to look at the question more broadly. This one comes from a paper presented at a Cornell University conference by University of California–Davis animal sciences professor Frank Mitloehner.[34] He points out that FAO calculations fail to account for what he terms "default emissions." "[I]f domesticated livestock were reduced or even eliminated regionally, the question of what 'substitute' [greenhouse gases] would be produced in their place has never been estimated." In other words, if livestock were removed, would all the climate change emissions attributed to them (now 14 percent) be eliminated? Clearly not. For instance, the earlier discussion of Brazilian soy makes clear that foods eaten in place of meat will have environmental impacts of their own. Yet there's no sense in talking about mitigating climate change by reducing or getting rid of a particular source of emissions unless it can likewise be demonstrated that what replaces the source would have lower emissions.

Cattle provide much more than meat and milk. As living animals, they give invaluable organic fertilizer in the form of manure. In many parts of the world cattle also provide irreplaceable power for plowing and transport. Post-slaughter, their hides provide leather, while additional fertilizer is made from their blood and bones. And it goes far beyond that. Modern slaughterhouses and processing plants are exceptionally efficient at putting each part of the animal to use, making everything from pharmaceuticals to blood vessels used in human transplant procedures to tennis racquet strings. "Therefore, to estimate accurately the 'footprint' of all livestock 'default' emissions for non-livestock substitutes need to be estimated and compared to livestock emissions (e.g., manure versus fertilizer, leather versus vinyl, wool versus microfiber, etc.)," Mitloehner points out. Only then can "the net [greenhouse gas] differences between livestock and other land-use forms" be estimated and an accurate accounting of livestock's true climate change impact be calculated.

I've lingered a bit here on the calculations for both the 18 percent and the 14 percent numbers because the 18 percent figure has been so widely cited. My point here is not to argue for any particular figure, but rather to demonstrate that the matter of meat's connection to climate change, especially for cattle, is far from a "case closed." What's much more important than any precise figure is the overall question of whether or not cattle truly aggravate the world's very real and urgent global warming crisis.

And there are good reasons to doubt that they actually do, or at least that they *unavoidably* do.

Cattle as Climate Change Mitigators

In an extremely significant omission, carbon sequestration was entirely left out of both of FAO's calculations for meat's global warming contribution. Carbon sequestration is a way of taking carbon from the atmosphere and tucking it into the soil, improving soil's functionality along the way. Methods for enriching the carbon content of global soils include "working the land with the goal of building topsoil, encouraging the growth of deep-rooted plants, and increasing biodiversity," explains author Judith Schwartz. "[R]ather than focusing on growing crops, the intention is to grow the soil."[35]

FAO has noted its failure to take into account carbon sequestration and even acknowledged its significance. FAO justifies this decision by stating it "could not estimate changes in soil carbon stocks under current land use management practices because of the lack of global data bases and models."[36] In other words, FAO is saying that it's hard to accurately quantify carbon sequestration at this moment. So rather than include an estimate, it simply omitted sequestration entirely from the calculations. At the same time, however, the 2013 report acknowledges that "[g]rassland carbon sequestration could *significantly offset* emissions."[37] This suggests that even FAO's lower figure overstates cattle's role in global warming, perhaps considerably.

FAO's somewhat dismissive treatment of the issue might give the impression that carbon sequestration in soil is just the wishful thinking of creative cattle ranchers. Nothing could be farther from the truth. "Soil carbon sequestration at a global scale is considered the mechanism responsible for the greatest mitigation potential within the agricultural sector, with an estimated 90 percent contribution to the potential of what is technically feasible," states a 2012 study by an international team of scientists in the *Proceedings of the National Academy of Sciences*.[38] This number has gained broad acceptance among scientists. Virtually the same figure was used by scientific advisers to the IPCC, the international body set up by the United Nations Environment Program, which stated that "89% of agriculture's [greenhouse gas] mitigation potential resides in improving soil carbon levels."[39]

Despite its enormous potential for climate change mitigation, policy makers have for years ignored opportunities to capture (re-capture, actually) carbon in soils. In 2009, the UK Soil Association sought to address that neglect by undertaking the largest and most comprehensive scientific review about soil carbon ever done. Its scientists reviewed 39 comparative studies of organic farming soil carbon levels, from various countries and climates.

The Soil Association's review concludes that soils would achieve "high carbon gains" from conversion to organic agriculture. The 39 studies showed that farming that includes use of animal manures, crop rotations, cover cropping, and composting results in greater soil carbon. More specifically, the report found that organic farming enhances soil carbon levels by putting more organic matter back in soils, and in forms that efficiently produce soil carbon. This is achieved by integrating cropping and livestock systems, which assists formation of soil carbon, and by increasing the proportion of land covered in vegetation, which promotes microorganisms that stabilize soil carbon once formed.[40]

Farm animals, especially grazing animals, the Soil Association report notes, are essential to higher-carbon-sequestering farming. "Grass-fed livestock has a critical role to play in minimising carbon emissions from farming and this must be set against the methane emissions from cattle and sheep." Grasslands for livestock grazing, including permanent pastures and temporary grass leys (grass cover crops on mixed farms), hold "vitally important soil carbon stores," the study concludes. It also reports that plowing up grazing areas for crop production has resulted in the UK losing 1.6 million tons of carbon to the atmosphere every year (amounting to an additional 12 percent of the UK's agricultural greenhouse gas emissions).

Globally, transitioning to organic farming would lead to substantial carbon sequestration, the report concluded. With "organic best practices" it estimated a staggering potential for sequestering 1.5 billion tons of carbon per year, an offset of about 11 percent of *all* anthropogenic global greenhouse gases (not just those related to agriculture), for at least the next 20 years. And these calculations were intentionally conservative. Not included, for example, was "the increase in agricultural soil carbon storage that would result from the almost certainly greater percentage of farmland that would be in permanent grass with widespread organic farming."

Given its promise, you may be wondering why we hear so little about soil carbon in policy discussions and in the mainstream media's climate change stories. Could it be that soil—otherwise known as dirt—is simply not sexy enough to capture the public's imagination?

Soil erosion, arguably an even greater threat to the planet than climate change, has long suffered from the same fate. The subject may not be trendy or catchy, but soil is the earth's workhorse.

"Soil truly is the skin of the earth—the frontier between geology and biology,"[41] says earth and space sciences professor David Montgomery. "Soil is our most underappreciated, least valued, and yet essential natural resource."[42]

Not only is soil essential to food production, it also contains massive amounts of carbon. Carbon losses from the earth's soils have accounted for a full one-tenth of *all* human-caused carbon emissions since 1850. The carbon that remains in soils is still about *three times* as much as what's in the atmosphere and *five times* as much as what's contained in the world's forests. Put another way, more carbon resides in soil than in the atmosphere and all plant life combined, an estimated 2.5 trillion tons of carbon in soil, compared with 800 billion tons in the atmosphere and 560 billion tons in plant and animal life.[43] Even now, soils are such gigantic carbon banks that just a small percentage increase would have dramatic effects. For example, it is estimated that increasing average soil carbon levels by just 1 percent would reduce atmospheric CO_2 by 2 percent.[44]

The tantalizing part of this story is that, in contrast with emissions from burned fossil fuels, carbon losses from soils are reversible: The soil carbon bank can be re-created. "The principal component of the soil carbon store is humus," the Soil Association explains, "a stable form of organic carbon with an average life-time of hundreds to thousands of years." Soil humus levels also determine how much water soils hold and how well they drain. "Low soil carbon levels are therefore likely to exacerbate the impacts of climate change, by increasing the risk and severity of droughts, water shortages and surface-water flooding" the Soil Association notes.[45]

The science of soil carbon sequestration is, admittedly, still nascent. But it is far more sophisticated today than just a decade ago. Once, it was assumed that the amount of carbon in soils was determined simply by how much organic matter was added. Now it is known that complex questions of biology affect whether organic matter is converted to stable soil carbon; the amount can vary from just a few percent to up to 60 percent.[46]

At the risk of glazing over the reader's eyes, because this is such a crucial point, I'm going to get granular here for a moment about how soils store carbon. I promise to be quick about it. This is more scintillating than it sounds.

Here's an intriguing place to start: Soil sequestration takes glue, glue that literally tacks carbon to the earth, keeping it safe from washing or floating away. The sticky substance involved, a protein that under a microscope looks like strands of honey and is produced by underground fungi wrapped around plant roots, is called glomalin. In 1996 glomalin was discovered and named by USDA soil scientist Dr. Sara Wright, assisted by Kristine Nichols, who has carried the work forward to this day. Years of experi-mentation and field research have helped Drs. Wright and Nichols begin to unravel glomalin's complex functions as well as document that it makes up 15 to 20 percent of soil's organic matter.

Arbuscular mycorrhizal fungi, the fungi involved in glomalin synthesis, function symbiotically with plants. Formed as networks of fine threads (hyphae), they envelop plant roots and extend their thin fingers out into soils. These fungal filaments engage in two-way exchanges: transporting nutrients from the soil to the plant while carrying carbon from the plant to the soil. Glomalin is a carbon-based molecule, while mycorrhizal fungi are carbon-based life-forms. Both utterly depend on carbon, which they obtain only from the roots of a living plant. Nichols refers to carbon as the currency in these exchanges. The plant uses carbon to "buy" nutrients from the mycorrhizal fungus. The fungus may then use that carbon to "buy" nutrients from other microorganisms in the soil.

Glomalin "glue" plays several roles in the process. It coats the fungus's hyphal threads, assisting in nutrient cycling while also protecting them. Moreover, it aids in forming and stabilizing clumps, known as soil aggregates, helping plant health and soil structure.[47] (Proteins left behind by earthworm locomotion may also play a role in soil aggregation, although studies are somewhat conflicting.[48])

Formation of soil clumps may sound banal, but it is considered a cornerstone of carbon sequestration. In aggregation, soil's mineral particles are clustered together and encapsulated. This stabilizes the carbon-containing decomposed plant and animal matter (humus), and protects it from degradation. Carbon from plant roots has been shown to be especially stable, lasting over twice as long in the soil as carbon from plant stems and leaves.[49] Soil aggregation gets carbon into a stable situation and keeps it there.

Healthy soil is filled with these tiny clumps. Soil aggregates "provide structure to soil for better water infiltration, water holding capacity, and gas exchange, and increase soil fertility by providing organic carbon (that is, food) to soil organisms, which use this food as energy to release plant nutrients from the soil," Nichols explains.[50]

Nichols's ongoing field research shows that soils under native grasses in North Dakota—for example, switchgrass, blue grama, big bluestem, and Indian grass—have higher levels of glomalin than those planted with non-native grasses. This raises the intriguing possibility that restoring native grasslands could be a key component of a global climate change mitigation strategy. "The more glomalin in a particular soil, the better that soil probably is," says Nichols. And up to a certain saturation point, the more glomalin, the more potential for carbon sequestration.

Correspondingly, Dr. Nichols's research has shown that plowing ground injures the root-loving fungi and reduces glomalin. The living hyphal networks "are physically torn by tillage," Nichols explains. When tillage disturbs soils, the fungi must allocate carbon to rebuilding their filamentous networks rather than either extending the networks or producing glomalin. In contrast, farming with low or no tillage increases both mycorrhizal fungi and glomalin, with levels particularly robust in rangelands and in cropping systems with diverse rotations.[51]

This suggests that the greatest opportunities for carbon sequestration lie in grazing areas (rangelands and pasturelands), especially those with native grasses, as well as in diversified operations where grass is part of a multicrop rotation.

To effectively serve as the indispensable mediator in these underground exchanges, the fungus must maintain an extensive hyphal network: The greater the fungus's network, the more interactions it can have with bacteria and other microbes inhabiting the soil. Those microorganisms decompose organic matter and make the nutrients available for the kind of carbon-based bartering I've just described. The transactions can be quite convoluted. For example, the fungus may feed bacteria growing on its hyphae some of the carbon it gets from the plant. In exchange, the bacteria use this carbon to produce enzymes, which will then be used to make nutrients available for the plant.

Having diverse crop rotations aids the complex system by enhancing the variety of organisms living in the soil. This, in turn, improves nutrient cycling as decomposition and enzymatic reactions release bound-up minerals. The fungi carry these minerals to the plant and exchange them for carbon.

Nichols has learned that using synthetic fertilizers, particularly phosphorus, creates unfavorable conditions for such transactions.

Phosphorus is a key nutrient used by mycorrhizal fungi in trade for carbon from the plant. When commercial fertilizer is added, the plant simply takes nutrients from the fertilizer rather than "buying" those nutrients from the fungus in exchange for carbon. Synthetic nutrients ultimately lead to debilitated mycorrhizal networks and lower levels of glomalin.

As you can tell, there is infinitely more going on beneath our feet than most of us have ever imagined. Keeping living plants in the ground for as much of the year as possible is also critical to powering this bustling underground economy, Nichols explained to me by email. The Soil Association report likewise emphasizes that the more time vegetation covers the soils, the better.[52] Growing plants conduct photosynthesis. Photosynthesis provides the carbon needed for mycorrhizal fungal growth and glomalin production. Without plants in the ground, the cascade of intricate processes grinds to a halt.

The newly emerging science of glomalin offers further evidence of the importance of grazing animals to our food system. Equally important, it reinforces the pivotal role they would play in a well-designed food system that has environmental sustainability as a primary objective. To maximize glomalin in soils, we need grazing lands. No agricultural lands are more continuously covered with vegetation than well-managed permanent pasture and rangelands. Properly stewarding land to raise healthy cattle herds also means managing it for vibrant underground communities, which in turn fosters the capture and storage of atmospheric carbon. What's more, grass is uniquely capable of forming soil carbon, as the UK Soil Association's carbon sequestration report points out. Grass has high root densities, fine root hairs, and high mycorrhizal fungal levels, all of which increase soil aggregation. Animal manures and composts also promote soil carbon, again reinforcing how including animals in farming benefits carbon sequestration.[53]

Cattle grazing not only ensures the existence of grasslands, and makes protecting open space from development economically viable for landowners, it actually aids in keeping those lands well covered in growing vegetation. Properly timed cattle grazing can increase vegetation by as much as 45 percent, North Dakota State University researchers have found.[54] And grazing by large herbivores (cattle included) is essential for well-functioning prairie ecosystems, research at Kansas State University has determined.[55]

The Kansas-based nonprofit Land Institute agrees. The institute has presented the Obama administration with a 50-Year Farm Bill that proposes increases in perennial crops and permanent pasture. "We see future herbaceous perennial grain-producing polycultures being managed through fire and grazing, just as the native prairie was 'managed,'" institute president Wes Jackson told me. "The large grazer on grassland has always been an integral part of the system here in North America."[56]

Not surprisingly, converting grassland to cropland causes carbon losses, while converting cropland to grass results in carbon gains. Soil experts Rattan Lal and B. A. Stewart collaborated in a major survey and report, with recommendations, regarding the world's soils. Lal is a professor of soil physics in the School of Natural Resources and director of the Carbon Management and Sequestration Center, Food, Agricultural, and Environmental Services, at Ohio State University. Stewart is a distinguished professor of soil science at the West Texas A&M University, and is a past president of the Soil Science Society of America.

In "Food Security and Soil Quality," Lal and Stewart report that "the potential of fallows to increase carbon contents has also been widely documented."[57] They note that due to the destructive effects of tillage on populations of "soil macrofauna" (like insects), soil cultivation "can strongly modify the soil functioning and especially soil carbon storage."[58] They report an especially notable decline in soil organic carbon when natural vegetation is replaced with crops.[59] Likewise, a study published in the journal *Global Change Biology*, which analyzed 74 soil carbon research studies, showed that merely converting cropland to pasture resulted in a 19 percent increase in soil carbon levels.[60] Animal grazing, as we have seen, maintains grassland ecosystems both above- and belowground.

Skeptics of soil sequestration's potential argue that the rates of sequestration tend to diminish within a couple of decades after a switch from cropland to pasture or from improved farming practices. However, as the Soil Association points out, "it is the next 20 years that will be critical in policy terms for delivering major greenhouse gas reductions." Moreover, they note, carbon sequestration continues for as much as 100 years, just at lower rates.[61]

From Dr. Lal's many years of research into soil carbon, he believes that enormous amounts of carbon could be captured in soils. Lal has concluded that carbon in the world's agricultural soils has been depleted between 50 to 70 percent. But he sees potential for restoration efforts to increase the carbon content of the globe's soils by upward of one to three billion tons a year, equivalent to approximately 3.5 to 11 billion tons of CO_2 emissions[62]—in other words, as much as one-third of all annual human-generated carbon emissions.[63]

In her recent book Cows Save the Planet, author Judith Schwartz explores the work of soil scientists like Dr. Lal as well as ranching practitioners, making a credible and compelling case that carbon sequestration triggered by well-managed grazing could be part of an effective strategy to combat climate change. Among scientists advocating for grazing animals as a means to foster carbon sequestration is Australian soil ecologist Dr. Christine Jones. Her blueprint for building carbon-rich soils includes short pulses of grazing with high densities of animals. With such land management, Jones says, "evidence of new topsoil formation can be seen within 12 months, with quite dramatic effects often observed within three years."[64]

This still leaves the question of soil carbon sequestration's potential magnitude. Is it limited to mitigating some of the greenhouse gases from cattle or does it raise the tantalizing prospect that cattle could actually have a net *positive* impact on climate change? And this is where things really get interesting.

To this point, I've been arguing merely that cattle are blamed for a disproportionate share of greenhouse gases, and that those gases can be largely offset by multiple beneficial effects of grazing animals in the farming system. Now I want to propose a far more radical idea: Cattle are not, in fact, a climate change *problem* at all; instead, cattle are actually among the most practical, cost-effective *solutions* to the warming of the planet. Unthinkable? Perhaps. But this is the very case currently being made on the world stage by wildlife ecologist Allan Savory.

Although not himself a rancher, Savory has been well known for decades in farming and ranching circles for his innovative teachings on planning and grazing, a system he calls Holistic Management Grazing or Holistic Planned Grazing. In recent years, Allan Savory's fame has been going mainstream. This is largely thanks to a March 2013 TED talk (viewed by 2.5 million people, at last check) in which he takes the controversial position that cattle are the world's single best hope for reversing climate change.

Savory is a true iconoclast, equally offending nearly anyone imaginable: academics, ranchers, conservationists, and policy makers, among others. His advocacy for using the bovine, as he puts it, "as a bulldozer that doesn't use diesel, to improve the land" particularly outrages the legions of environmental and vegan advocates who want to see cattle raising go the way of the butter churn. Meanwhile, his ideas that herds should be dense, not sparse, that compaction is beneficial to the land, and that "native" and "non-native" plant designations are practically mean-ingless, run directly counter to mainstream rangeland and ecology teachings. He hates feedlots and disagrees vehemently with the way most cattle are being raised in the industrialized world today, so neither the cattle nor the beef sector has much use for him. He has, in other words, no built-in constituency.

Yet Savory is simply too credible to be ignored. For decades, he has tirelessly labored to restore grasslands on five continents. In the United States, he founded and leads the Savory Institute, which works both through seminars and publications and out on the land, teaching and demonstrating land improvement through well-planned grazing. He's given major lectures at leading academic institutions including Harvard, UC Berkeley, and Tufts universities. The Savory Institute and he himself have received highly prestigious awards, including the 2010 Buckminster Fuller Challenge award for the organization's work "to solve the world's most pressing problems," and the 2003 Banksia International Award, given to the person doing the most for the environment on a global scale.

Today, the Savory methods are being followed in the management of some 40 million acres the world over.[65] Some of the results are breathtaking. I have been on several ranches (including our own) that are generally following Savory's ideas, and have seen photographs from locations in several parts of the world where his ideas are being put more strictly into practice. Arid, denuded lands once classified by ecologists as "beyond restoration" have been transformed into areas rich in water, and teeming with plants and animals. Biodiversity has soared.

And astonishingly, cattle are the linchpin.

How can that even be possible? We all "know" that big, heavy animals and, cattle above all, do not improve land, they damage it. They do not add water or vegetation to the land, they diminish it. And surely they don't benefit other animal populations, they decimate them, right?

For the first part of his life, Allan Savory says he "knew" all these things, too, and he knew them well. As a youngster growing up in southern Africa, he was often in ecologically damaged areas that were said to be stressed by too many animals, wild or domesticated. As a university student of wildlife ecology, Savory was taught again and again the incontrovertible truth that soil erosion and desertification around the globe resulted from animals trampling and chopping up land, and, especially, from them grazing too heavily on the vegetation. "I was taught that overgrazing was causing the land to turn to deserts," he says. "We were once equally certain that the world was flat."[66]

"My education began after I left university," he has also said.[67] But when he started his career, he accepted what the textbooks had taught him. So certain was he of these doctrines that when he became a government game officer in Northern Rhodesia (now Zimbabwe), most of his work to restore and protect wild areas involved getting animals off the land. "Resting" land by removing large animals was considered the surest and best land salvage strategy.

As a young wildlife ecologist employed to protect an important game preserve, Savory wrote a report that recommended killing 30,000 elephants. "It was the hardest thing I ever did," he recalls, "because I love elephants." The idea of shooting tens of thousands of elephants was not only repugnant to Savory, but also horribly unpopular with the public. For this reason, a peer review of the report by other ecologists was ordered. The reviewing scientists agreed: To restore the land, reducing elephant numbers was essential.

Because of Savory's report, some 20,000 elephants were killed. As horrific as that was, the worst was yet to come. That was when it became apparent that despite the slaughter, the land's condition was worsening, not improving. Killing tens of thousands of elephants had all been for nothing. Savory now refers to his recommendation to cull all those elephants as the biggest mistake of his life.

"We all believed," Savory wrote in 1988 about his earlier work in African game reserves, "that with dramatically fewer animals living there the area would naturally recover, but it continued to deteriorate." Like all his colleagues, where cattle were present, he blamed them. "In my days as a young scientist, I hated livestock and the people running them,"[68] he recounts.

Actually, Savory says, he later realized that "[w]e scientists had the bull by the udder."[69] Again and again, Savory was witnessing reductions in animal numbers resulting in greater environmental degradation.

Savory says one good thing did come of these tragic experiences. They set him on the path of dedicating his life to reversing desertification through practices that were actually working in the field. Through study and observation, he eventually came to believe that animal impacts are key to properly functioning ecosystems. Winnowing down their numbers is not the answer, not in the least. Instead, the solution lies in making herds of domesticated animals function more like the wild herds of herbivores with which the ecosystems evolved.

"We can now see that it was not the livestock per se that caused significant damage in the past, but the way we managed them,"[70] Savory says. He decries widespread damage in many parts of the globe from improper livestock management. Yet he now believes that more often than not land in the American West and elsewhere is not being overgrazed. Rather, it is, in Savory's words, being "over-rested." Although that statement runs strongly counter to conventional wisdom, the critical importance that *disturbance* on the landscape plays in the diversity, structure, and function of ecological systems has been borne out by numerous scientific studies.[71]

Savory believes desertification results primarily from reduced biodiversity—the loss of the mass and diversity of plant and animal life. He notes that sustaining life in any environment requires maintenance of a basic life cycle of birth, growth, death, and decay, in which nutrients are cycled continuously. In perennially humid environments, decay is fostered by a high population of microorganisms. "As vegetation dies throughout the year, organisms continually break it down."[72]

For areas with less moisture, the cycling of life is more problematic. Savory believes that for those parts of the world, the question is how "brittle" they are. Brittleness is a concept distinct from fragility. "[D]etermining the degree of brittleness becomes a prime factor in the management of any environment," Savory wrote in his 1988 book *Holistic Resource Management.*[73] The brittleness of an environment is based on how organic matter decays, and on the order of succession of organisms in the environment.[74] The least brittle environments have 100 percent biological decay processes, whereas chemical (oxidative) and physical (weathering) decay processes become dominant as one moves toward brittleness.[75]

Moisture in the environment is a critical factor. But much more important than total rainfall is "the distribution of precipitation and atmospheric humidity throughout the year," Savory writes.[76] Other factors affecting how brittle an environment is include elevation, temperature, and prevailing winds. "Brittle environments commonly have a long period of non-growth, which can be very arid."[77]

Understanding an area's brittleness is important, according to Savory, because it determines the way land should be managed, which is highly site-specific. "[T]he old belief that all land should be left undisturbed in order to reverse its deterioration has proven wrong," he wrote. "Only non-brittle environments respond this way. In brittle environments, prolonged non-disturbance will lead to further deterioration and instability."[78] Grasslands, he found, could lie anywhere from 1 to 10 on the scale of brittleness.[79]

The most important indicator of grassland health is the amount of bare ground in between plants, according to Savory. The more bare ground, the worse the land's health. Ground without vegetation is exposed to the force of wind and the power of rainfall, which blow and wash soils away.

Nowadays, Savory emphasizes what he terms *effective rain*. Total rainfall is irrelevant, he says; what matters is rainwater taken up by plants, held in soils, or going into groundwater. *Ineffective rain*, in contrast, is that rainwater that runs off or evaporates.[80] Soils, he notes, are by far the world's greatest reservoir of fresh water.[81]

I first heard about Savory and his unique methods for land restoration more than a decade ago, from people who had worked at our own ranch and taken courses on grazing management. His books have always been on our own bookshelves, and his teachings have long served as the basis for how our own ranch is managed. Keeping cattle fairly densely congregated, in prescribed tracts of land, moving them frequently, and allowing the land to rest for extended periods between grazings are all ideas adapted from Savory's teachings.

Yet Savory advocates a much more particular and comprehensive management approach than the one we currently follow. And it's an aspiration of mine that our ranch will continue, over time, to move still closer toward his methods.

Stated succinctly, Savory advocates that animals be kept in dense herds and moved often; that grazing stimulates biological activity in the soil; that animal waste adds fertility; that hooves break the soil surface, press in seeds, and push down dead plant matter so it can be acted upon by soil microorganisms; that all of this generates soil carbon, plant carbon, and water retention; and that this is the only way to stop and to reverse desertification the world over.[82] "The actual grazing plan will thus be different on each ranch (in each season) and will be continuously changing, according to conditions on the ground."[83]

The basis for Savory's approach is to re-create, to the extent possible, the conditions under which grasslands evolved. He does not argue that cattle don't alter the land. Quite the reverse. After a long career working as a wildlife ecologist and land restoration adviser, Savory is the first to acknowledge that the presence of cattle changes a landscape's ecology. "Where a cow places her hoof today begins a chain of reactions that ensures that that spot will never be exactly the same again," he states.[84] He classifies the bovine's impacts on the land into three types: compaction ("heavy animals on small hooves"); breaking the land surface; and cycling vegetation faster than if they weren't there.[85]

When cattle are not well managed, these impacts can be extremely damaging. He notes that in most arid and semi-arid regions humans have long grazed livestock, and usually in ways that harm the land. "When livestock management practices produce bare ground, a critical share of available moisture either evaporates or runs off. Springs dry up; silt chokes dams, rivers, and irrigation ditches; and less water remains for agriculture, industry, and people in nearby cities."[86]

Many ecologists would agree with that statement, but they would attribute the damage to something else: overgrazing and the presence of too many domesticated grazing animals. Here Savory's views diverge radically from the mainstream. "Overgrazing has nothing to do with animal numbers and everything to do with timing," he urges. "Think of it like your lawn: it makes no difference to the grass if you use one mower or 50 mowers. All that matters is how frequently you mow it."[87]

In fact, Savory says, significant numbers of animals are absolutely essential, and they must be kept in dense herds. Sparse numbers are insufficient to cause desirable disturbances. Trampling and breaking the ground surface, as hooved animals are particularly prone to doing, Savory argues, are actually essential to grassland ecosystem function. Such was the impact that massive herds of wild herbivores (along with the predators that stalked them, whose presence ensured that herbivores stayed closely congregated) had on the earth for millions of years. And under such conditions grasslands evolved. Stated another way, the plants and microorganisms of grassland vegetation co-evolved with animals not just to *tolerate*, but to actually *need* grazing.

Here's how Savory himself explains it:

> [M]ost of these perennial grass plants developed alongside vast herding herbivores. The animals were dependent on the grass for food, and the grass was dependent on the animals for removal and rapid biological decay of the dying material. Most grasses have growth points close to the ground, out of harm from grazing animals, so the animals could remove leaves without damage to the plant and also allow sunlight to reach the growth points the following season. But if there are not enough large herbivores with vast microorganism populations in their moist gut to cycle this mass of material, then it remains standing.
>
> As dead grass parts stand, without large herbivores to bring about decay, a chemical process of oxidation takes over. Oxidation is a very slow process: experimental plots show that it can take over 60 years for a perennial grass to finally break down if totally protected from fire or animals. Slow oxidation/weathering leads to premature death of most animal-dependent perennial grass plants, and thus to loss of biomass and diversity. Healthy grassland gives

way to woody tap-rooted plants that are not dependent on grazing herbivores (if rainfall is high enough). And in places where rainfall is low, it gives way to desert bushes and bare soil—i.e., desertification.[88]

As already discussed, grazing animals are absolutely essential to properly functioning ecosystems. A March 2014 grazing impacts study by the University of Maryland provides just one example. The researchers carried out field tests at sites on six continents, creating various adjacent test plots that either allowed or excluded grazing to specifically catalog grazing's effects. They consistently found greater plant diversity flourishing where animals were allowed to graze compared with plots where animals were excluded. The scientists concluded that because animal grazing allows more sunlight to reach the ground, more diverse types of plants were able to grow.[89]

In relatively recent history the age-old co-evolutionary cycle of animals and grasslands was severely ruptured. By about 10,000 years ago, human hunting had decimated native roaming animal herds over much of the globe. Today, mere remnants of those once mighty animal mobs linger. Wild grazing herds are sparse, and very few of the world's large predators remain. These anemic populations cannot and do not have a similar impact on grasslands. It is thus the *absence* of animals rather than their *presence* that has caused land to become increasingly arid and desertified in recent millennia. This is why Savory talks about "over-resting."

He argues further that the tools to affect large tracts of land, such as the world's grasslands, are limited. Humans have at their disposal the following: technology (machines, chemicals, and so forth); resting (which can occur while being grazed); fire; and grazing.[90]

Of these alternatives, Savory believes cattle offer the best hope for reestablishing balance. Managed correctly, cattle herds can carry out many of the ecosystem services once performed by wild herds. "We've got to *use* animals—their mouths to graze, their hooves to lay litter, to cover soil, to chip the surfaces, and more than anything else we're using the micro-organisms that are in their gut," Savory explains. "We are using the livestock as a proxy for the wild animals that were there before. . . . Nothing else in the world is available to do this except livestock."[91] He urges, "There is no other known tool available to humans with which to address desertification."[92]

What's particularly intriguing about Savory's ardent advocacy for cattle grazing is that it comes from an ecologist, not a rancher; he has, he says, never been passionate about livestock. Instead, he has a lifelong, unwavering passion for wildlife. Only after years of working to restore game preserves did he realize that to protect wildlife, livestock were essential.[93]

Savory now lives mostly in the United States, but he remains strongly connected to his African roots. In those parts of the world where the desertification crisis is acute, Savory notes, expanding deserts are not only a climate change concern but also a major cause of conflict, poverty, and emigration.[94] At the Africa Centre for Holistic Management in Victoria Falls, Zimbabwe, a 200-hectare "learning site," Savory's methods are on full display. There, as they manage for benefits to wildlife, not livestock, they have increased the number of cattle year after year.[95] Cattle are indispensable, according to Savory. "We *need* cattle on the lands to stop and reverse desertification," he says.[96] "Leaving it to nature doesn't work because [the ecosystems] aren't natural anymore."[97]

Savory's global following is growing because he has demonstrated that what he is advocating works, and the results have been nothing short of astounding. On tracts that he directly manages as well as on the millions of acres where his teachings have been put into practice, land that had dried out and begun turning to desert has been restored to grasslands teeming with water and life. Although his climate change arguments run counter to mainstream thought, they are simply too compelling to be ignored.

Numerous empirical studies have documented multiple ecological benefits to using methods like those advocated by Savory. In one such study, researchers at Texas A&M University compared the effects of different grazing methods, including a multi-paddock grazing with high stocking density (*mob grazing*), on several ranches that had each used the same management methods for at least nine years. The research team found wide-ranging benefits to the mob grazing method, including the following: less bare ground; more soil aggregation; soil that is more penetrable; lower sediment loss; higher soil organic matter; and the highest fungal/bacterial ratio, "indicating superior water-holding capacity and nutrient availability and retention." Overall, the researchers concluded: "This study documents the positive results for long-term maintenance of resources and economic viability by ranchers who use adaptive management and [mob] grazing . . ."[98]

Similarly, a multiyear field study done by researchers at Idaho State University compared different land management strategies for their effects on soil-water content. The three strategies were simulated Holistic Planned Grazing (the Savory approach), rest-rotation (lighter stock density grazing), and total rest. Soil-water content was measured continuously for two years using 36 sensors. The researchers found both that the percentage of ground covered by plant litter was the highest and that the soil-water content was highest using the Savory method. "Management decisions (grazing and rest) can have substantial influence upon soil-water content and . . . soil-water content can vary substantially as a result of animal impact and the duration of grazing," the study concluded.[99]

Now the burning question many have been asking is whether the impact of Savory's methods upon climate change can be quantified. Fortunately, a credible effort to do just that has been under way.

Seth Itzkan is an engineer who heads a consulting firm called Planet-TECH Associates whose work includes land restoration and climate change mitigation. Itzkan became so intrigued by the potential for grazing as a way to fight climate change that he traveled to the holistic management center in Zimbabwe, staying for six weeks. On three separate occasions now, Itzkan has visited, each time for four to six weeks, to study how Savory's grazing methods impact land, water, and wildlife. Combining observations and field research in Zimbabwe with extant research on soil carbon capture, Itzkan published a 34-page technical paper in April 2014 calculating the effect that Savory's grazing methods could have globally for ameliorating climate change.[100]

Itzkan's report quantifies the potential for carbon capture of Savory's grazing strategies. He concluded that holistic grazing could sequester between 25 to 60 tons of carbon per hectare per year in semi-arid grasslands. As more and more carbon was captured, land would shift from semi-desert shrubland to healthy tropical savanna and perennial grassland. Itzkan also calculated that the total potential for capturing soil organic carbon in grasslands is around 88 to 210 gigatons, which translates to approximately 41 to 99 ppm atmospheric CO_2. This, the author notes, is "enough to dramatically mitigate global warming."

Cattle and Climate: The Final Word

Sometimes, I get discouraged by the oversimplification of these issues and the way beef gets vilified in climate change discussions. But I found it heartening when I recently read a Letter to the Editor at *Grist* from one of the world's best-known vegetable farmers, Eliot Coleman, who also happens to be a former vegetarian. Responding to a *Grist* article (one of the countless) about how meat bears a disproportionate share of the blame for global warming, Coleman wrote:

> When I think about the challenge of feeding northern New England, where I live, from our own resources, I cannot imagine being able to do that successfully without ruminant livestock able to convert the pasture grasses into food. It would not be either easy or wise to grow arable crops on the stony and/or hilly land that has served us for so long as productive pasture. By comparison with my grass fed steer, the soybeans cultivated for a vegetarian's dinner, if done with motorized equipment, are responsible for increased CO_2 . . . . Targeting livestock as a smoke screen in the climate

change controversy is a very mistaken path to take since it results in hiding our inability to deal with the real causes [burning of fossil fuels].[101]

Like nearly all agricultural practitioners I've met, Coleman certainly understands full well that growing food is messy and complex, much more so than non-farmers tend to assume. Evidently he appreciates, as well, the unique and beneficial role of ruminants, especially if farming is to be done in a regenerative way rather than based on finite resources. I particularly appreciated Coleman's message that focusing on emissions from cattle is a red herring that takes attention away from where it should really be: fossil fuels.

I'm not really sure there's a good reason to quantify cattle's greenhouse gas emissions, but here's my take. I think FAO numbers show a worst-case scenario, especially for cattle, because the report is intended to support policies favoring a shift toward non-grazing animals. I find the cattle emissions figures greatly inflated, especially by including deforestation from developing countries, which is unfair and unreasonable. I also think FAO's numbers, especially for cattle, should be taken with a giant grain of salt because of carbon sequestration, which may be more than enough to completely offset the emissions from grazing animals.

That said, since FAO is the most credible entity doing such calculations, I use its numbers to roughly calculate the contribution of all cattle at 8.5 percent. Using FAO's current estimate of 14 percent for all animal-based foods, and, using a "CO_2 equivalence" for beef and milk production from cattle respectively of 41 and 20 percent of all livestock sector emissions (also from the report)[102]— that means foods from cattle are responsible for 61 percent of the 14 percent, in other words, 8.5 percent of greenhouse gases.

At the same time, FAO's report also shows that the emissions from cattle are not at all unavoidable. It notes that "emissions from energy consumption on farms and in processing are negligible in beef and limited in dairy" (about 8 percent of the sector's entire emissions).[103] And it establishes the enormous portion attributable to animal feeds. "Feed production, processing, and feed makes up the entire sector's biggest share, at 45 percent."[104] This huge figure at once reminds us of the special value of grazing animals and clarifies the urgency of getting cattle back onto grass.

Moreover, FAO acknowledges that beef and dairy emissions can be partially, and perhaps entirely, offset. Getting cattle onto grass is one way. The other major way is through carbon sequestration connected to livestock grazing, which nearly everyone agrees can substantially offset the sector's emissions. When the offsets are taken into consideration, the greenhouse gas emissions from cattle may be very small, or may be zero, or may even be a negative number. Whatever the number, it needs to be considered in context alongside the tremendous benefits of having cattle as part of our food system.

Soil scientists Lal and Stewart point out that there are two possible approaches to mitigating climate change. First, there's limiting greenhouse gas emissions. Second, there's improving the uptake of greenhouse gases from the atmosphere into stable pools.[105] As this chapter shows, soils hold enormous potential for the second part of the approach. And cattle are essential to the strategy.

Lal and Stewart argue that even beyond climate change, improving the world's soils is "an important and necessary step to free much of humanity from perpetual poverty, malnutrition, hunger, and substandard living." The key, they believe, is focusing on soil's physical and biological fertility. "Conservation agriculture that combines reduced tillage, crop residue retention, and functional crop rotations, together with adequate crop and system management, permit the adequate productivity, stability, and sustainability of agriculture," they say.[106] Chemical fertility, in other words, will not do. In May 2014, when interviewed by *The Boston Globe*, Lal said he considers Allan Savory's method, using carefully planned mob grazing, an effective way to increase soil carbon content.[107]

What's certain is that efforts to minimize greenhouse gases must be far more sophisticated than blanket condemnations of beef. Studies clearly establish that varying methods of animal husbandry lead to highly variable global warming contributions, or even mitigations. This variability is, in fact, explicitly acknowledged in FAO's most recent climate change report, which states:

> For ruminant products especially, but also for pork and chicken meat and eggs, emission intensities vary greatly among producers. Different agro-ecological conditions, farming practices and supply chain management explain this heterogeneity, observed both within and across production systems. It is within this variability—or gap between producers with highest emission intensity and those with lowest emission intensity—that many mitigation options can be found.[108]

Moreover, the same type of variability is true for every food we eat, not just for meat. A study by Sweden's national environmental agency showed that, depending on how and where a vegetable is produced, its carbon dioxide emissions vary by a factor of 10.[109]

At the same time, it should always be kept in mind that what happens on the farm or ranch is only a portion of the story for greenhouse gas emissions in the food system. Only about one-fifth of the food system's energy use is farm-related, according to University of Wisconsin research.[110] And the UK Soil Association estimates that in Britain, only half of food's total greenhouse impact has any connection to farms.[111] The rest comes from processing, transportation, storage, retailing, and food preparation.[112] The seemingly innocent potato chip, for instance, turns out to be a dreadfully climate-hostile food, mostly because of what happens after the potato leaves the farm.[113] Foods that are minimally processed, in season, and locally grown, like those available at farmers' markets and backyard gardens, are generally the most climate-friendly.

Rampant waste at the processing, retail, and household stages compounds the problem. Almost half of the food produced in the United States is thrown away, according to University of Arizona research.[114] Thus, a consumer could measurably reduce personal global warming impact simply by more judicious grocery purchasing and use.

None of us, whether we are vegan or omnivore, can avoid eating foods that play a role in global warming. Because the effects of any given production system are so variable, and because cattle grazing offers enormous potential for climate change mitigation, singling out beef is both unhelpful and misleading.

Instead, consumers should be encouraged to eat in ways that support environmentally sound, healthful food, and carbon sequestration. Every able-bodied person with a yard or a terrace should grow a portion of her own food. After Americans were urged to plant "Victory Gardens" during World War II, it is said they collectively grew about 40 percent of the vegetables they ate.[115] And because it generally takes more resources to produce meat and dairy than, say, fresh locally grown carrots, it's sensible to cut back on consumption of animal-based foods. As I wrote in *Righteous Porkchop*, we should eat less meat, eat *better* meat.

Eaters can also lower their personal global warming contributions by following these simple rules: Avoid processed foods and those from industrialized farms; reduce food waste; and buy local and in season.

Regardless of the components of our diets, it is absurd to focus narrowly on climate change to the exclusion of all else. As this entire discussion has hopefully made clear, every part of the agricultural system—whether raising carrots, soy, or beef—has multiple and varied effects on humans, animals, and the environment. I wholeheartedly embrace the idea that our food choices matter. We should choose wisely, with consideration not just of taste but also health and of the many ripple effects of the systems we are supporting. However, I soundly reject the suggestion that our ethical obligations start and end with climate change impacts.

Climate change is important and urgent, but it is one of many vital issues related to the health of our planet. Other parts of this book are dedicated to examining various environmental issues and social issues, well beyond climate change. For each of us individually as well as collectively, climate change effects should be one of myriad factors given consideration. As a society, we must plan food systems that create healthful foods while being ecologically sound and regenerative, as well as humane to workers and animals. As eaters, we should try to select ethically produced foods that are also healthy and delicious.

As with the production of all foods (and all consumer goods), raising cattle and turning them into beef has many and varied impacts—both positive and negative. The only sensible way to evaluate beef is through consideration of all of these effects.

Finally, if you wish to shape the food system with your food choices, quitting beef would have far less impact than shifting from commercial beef to well-raised beef. The global market for commodity beef is so massive that the purchasing dollars of any individual are like a drop of water in the ocean. In contrast, when you buy grass-fed beef or well-raised grain-finished beef from a farm or ranch in your region, you may *literally* be providing the financial support that means the difference between the survival or failure of that family business.

How, then, should we regard the role of cattle in climate change? Clearly the issue is complex, far more so than bumper stickers or FAO press releases would have us believe. But if we want a slogan, perhaps instead of equating hamburgers and Hummers, it should be: *Beef Puts Carbon Back Where It Belongs.*

CHAPTER 2

ALL FOOD IS GRASS

MODERN URBAN DWELLERS TEND TO LOOK at grass and see only monotonous, ubiquitous, inedible green ground cover. Environmentally minded landscapers loathe it and have taught us skepticism of expansive, watered, suburban lawns. These days we don't regard grass as part of our food system, let alone its foundation. Grass's connection to what we eat seems remote, if we think of it at all.

Why should a book on beef dedicate so many pages to grass? Evidently, cattle eat it. But that's just the beginning of the story. In the previous chapter, I showed how leaves and roots of grass, working with fungi and other soil microorganisms, as well as whole grassland ecosystems all play critical roles in staving off the warming of our planet. And despite our general lack of appreciation, grasses are the most important plants in the world. They are also inextricably linked to cattle.

Grasses cover around 40 percent of the earth's land surface, about 70 percent of the world's agricultural area,[1] and are the fourth-largest plant family in the world, containing more than 11,000 species worldwide.[2] As described earlier, the bovine has a special gift, a rumen that allows it to live off little more than grass—cellulosic vegetation that's inedible for people and most other animals. That unique ability of cattle to survive on widely available, naturally occurring vegetation without being provided feeds has made them essential to humans in many climates and geographies for thousands of years. Cattle have accompanied human migrations as they have fertilized our fields and gardens, powered our plows and carts, clothed our bodies with their hides, provided milk and meat—all while living primarily or entirely on a diet of grass.

Cornell University professor of agriculture and ecology David Pimentel is famously critical of the food industry, especially the meat sector. Yet his classic treatise *Food, Energy and Society* (co-authored with his wife, Marcia) repeatedly emphasizes the irreplaceable role of cattle and other grazing animals for human survival. "Cattle, sheep, and goats will continue to be of value because they convert grasses and shrubs on pastures and range-land into food suitable for humans. Without livestock, humans cannot make use of this type of vegetation on marginal lands."[3]

Grass is, in fact, the base layer of the global food system. It continually converts massive quantities of solar energy into food for grazing animals. Grasses and herbivores, working together, are the indispensable intermediaries between humans and the energy of the sun.

Their invaluable cohorts, of course, are water and soil. Earth and space sciences professor David Montgomery describes soil variously as "the dynamic interface between geology and biology, the bridge between the dead world of rock and the bustling realm of life,"[4] and "the thin layer of weathered rock, dead plants and animals, fungi, and microorganisms blanketing the planet," which "has been and always will be the mother of all terrestrial life." He considers it "every nation's most critical resource, one that is either renewable or not, depending on how it is used."[5] Soil, Montgomery says, connects "the rock that makes up our planet and the plants and animals that live off sunlight and nutrients leached out of rocks," which, "in the big picture . . . regulates the transfer of elements from inside the earth to the surrounding atmosphere."[6] Soil provides grass with essential nutrients while acting as the medium by which oxygen and water are supplied and retained. "Acting like a catalyst, good dirt allows plants to capture sunlight and convert solar energy and carbon dioxide into the carbohydrates that power terrestrial life right on up the food chain."[7]

Together, soil and water make grass's growth possible, while in return grass performs essential services for them. On average, 90 percent of grass's bulk is actually belowground, in the form of long, filamentous roots—a tangled network that stops soil erosion. "Grass prevents erosion by binding to the soil," explains the geology textbook *People and the Earth*. "Replacement of grasslands with cropland in the American Midwest, its 'breadbasket,' has caused nearly one-third of the topsoil to be stripped by erosion in the past 100 years."[8] Roots of grass tenaciously cling to soils, preventing them from being washed or blown away. Through complex biological processes, grass assists in forming new soils. By creating a massive web of tiny underground channels, and by enabling soil aggregation (as described in chapter 1), grass also renders the earth a far more effective keeper of water.

While even a casual observer can see that grasses feed wild and domestic herbivores, it may be less obvious that the relationship is two-way: Herbivores are nourished by grass, and in turn their grazing and trampling maintains grasslands. Research already presented makes clear that pruning by creatures' mouths supports vegetative growth. Grazing and trampling also keep down the emergence of woody plants that transform grasslands into environments less hospitable to grass. When grazing animals are extirpated or otherwise disappear from a terrain, the ecosystem, including all life depending on it, is radically altered. "When grazers are removed, grasses lose their competitive advantage and forbs and shrubs quickly become established," explains environmental resource management professor J. P. Curry.[9] Just as mowing helps keep a lawn grassy, lush, and full, so grazing triggers plant growth by pruning off old and dead parts and by exposing plants' growth points to more sunlight.

Beyond the lawn mower effect, grazing animals assist soils and grasses in other ways. Their hooves push seeds into the ground, preventing them from getting blown or washed away, or gobbled up by hungry birds. Cycles of decay, too, benefit from hooves pressing plant parts into soils. Down in the dirt, vegetation becomes enveloped by the decomposing microbes that power carbon and nutrient cycling. The presence of cattle also ensures that moist, nutrient-rich organic matter, long regarded by farmers as agricultural gold—and known to non-farmers as cow patties—is continually added to the soil. "Increasing soil organic matter by applying manure or similar materials can improve the water infiltration rate by as much as 150%."[10]

The cycling of vegetation through the bovine's unique, multi-chambered stomach hastens the grassland ecosystem's cycle of biological decay, supporting the soil's vast, complex underground economy.

Remember that topsoil harbors billions of microorganisms that help plants get nutrients from organic matter and mineral soil. Soil organisms "supply plants with nutrients by accelerating rock weathering and the decomposition of organic matter."[11]

Grazing of grasses thus bolsters continuous regeneration of soils and supports grass's diverse functions, from the microscopic to the ecosystem scale. Dense mats of sod shield soils from being blown away by wind and being washed away by water. Tangled grass root masses hold soils in place, and improve their ability to hold moisture.

"There are no better soil stabilizers than luxuriant pasture grasses and legumes, range grasses, and forest trees and shrubs," states the university textbook *Soil and Water Conservation*.[12] First, the text notes, because aboveground plants protect soils against the force of falling precipitation and sheeting water. "Close-growing perennial vegetation intercepts rain drops, thus reducing impact energy and decreasing soil dispersion and splash and sheet erosion. The velocity of surface water flow is reduced by contact with plant stems and residues." Second, because water soaks into the ground where grasses are growing, and soil particles are caught by grass blades rather than running off. "The result is more and faster infiltration of water into deep soil horizons, clean water flowing slowly along the soil surface, and less soil erosion and sediment." Finally, the text notes, plant roots cling to soil, protecting it from wind. "Vegetation also stabilizes soil against wind erosion by the 'holding' action of the plants roots and by decreasing wind velocity at the soil-atmosphere interface."[13]

A review of studies of world soils reported field trials that found nearly complete absorption of water (what Savory would call effective rain) with dense vegetative cover. By contrast, only 20 percent of rain infiltrated when the soil was bare.[14]

Even in seasons where they are dormant, grasses effectively safeguard soils. "Land areas covered by plant biomass, living or dead, are more protected and experience relatively little soil erosion because raindrop and wind energy are dissipated by the biomass layer and the topsoil is held by the biomass," according to the Pimentels.[15]

Wendell Berry contrasts a cornfield and prairie this way:

The most noticeable difference is that whereas the soil is washing away in the cornfield, it is building in the prairie. And there is another difference that explains that one: the corn is an annual, the cornfield is an annual monoculture, but the dominant feature of the native prairie sod is that it is composed of a balanced diversity of perennials: grasses, legumes, sunflowers, etc., etc. The prairie is self-renewing; it accumulates ecological capital; and by its own abounding fertility and diversity it controls pests and diseases. The agribusiness cornfield, on the other hand, is self-destructive; it consumes more ecological capital than it produces; and because it is monoculture, it invites pest and diseases.[16]

Naturally, the best protection to soils and water is provided by continuous vegetation—permanent pastures and rangelands. The Pimentels note: "Vegetative cover is the principal way to protect soil and water resources."[17] Wendell Berry writes: "If you want to stop soil erosion (so the prairie shows) you have to keep the ground covered—all the time, winter and summer. If you want to keep the ground covered all the time, the best way to do it is with a diverse, mutually beneficial polyculture of perennial plants."[18]

Compared with such grass-covered areas, "cropland is more susceptible to erosion because of frequent cultivation of the soils and the vegetation is often removed before crops are planted,"[19] *Food, Energy, and Society* notes. Fields used to grow crops are much more exposed to wind and rain energy because they are often partly or totally bare. Between plantings, the fields are nakedly exposed to the forces of wind and rain. When crops have been planted, seeds and seedlings can offer little protection. Even when crops are mature, "[r]ow crops are highly susceptible to erosion because the vegetation does not cover the entire soil surface."[20] On average, according to *Food, Energy, and Society*, erosion rates on U.S. pastures are 40 percent lower than on U.S. cropland.[21]

Other studies have reached even far more dramatic conclusions. A study by the environmental nonprofit Land Stewardship Project conducted field trials showing that compared with cropland, perennial pastures used for grazing can decrease soil erosion by over 80 percent.[22] Soil experts Lal and Stewart state: "Erosion rates from conventionally tilled agricultural fields average 1 to 2 orders of magnitude greater than erosion under native vegetation, and long-term geological erosion exceeds soil production."[23]

Even as part of a crop rotation, grasses, especially when grazed, provide a wealth of benefits. By delivering organic matter and carbon, as previously described, they feed the billions of tiny organisms teeming below the earth's surface. These microorganisms make nutrients bio-available, thus enhancing the earth's fertility. Grazing, by stimulating plant growth, and by cycling plant matter through the ruminant's gut, assists such carbon–nutrient exchanges. Healthy populations of soil microorganisms also provide protection from diseases, pests, and harmful fungi to crops grown in the same soils. Before there were chemical fertilizers, pesticides, and fungicides, there was grass.

Soils experts Lal and Stewart state that adding manure to the land enhances soil's chemical properties in several ways. First, manure is a source of carbon and nutrients, including nitrogen, phosphorus, and potassium. Second, manure improves soil's cation exchange capacity, its relative ability to store a variety of nutrients.[24] And manure stimulates beneficial biological activity. As previously noted with regard to phosphorus and glomalin production, chemical fertilizers function differently. "Mineral fertilization without organic amendments leads to the mineralization of [soil organic matter] and to a decrease in soil structure, pH, and the attendant decline in agronomic productivity," Lal and Stewart report.[25]

Its manifold benefits make grass the stalwart of sustainable global food production. Grassy areas with properly managed cattle herds also offer a stunning opportunity to sequester carbon, reverse desertification, and mitigate climate change, as previously detailed. Ideally, then, all food is grown with grass.

Earlier peoples knew little about such complex molecular interactions, yet they held the earth and grass in high esteem as sustenance for the animals on which they depended. "Out of earth you were taken, from soil you are and unto soil you shall return," says the book of Genesis.[26] The writings are replete with references to grass as life's foundation. The Psalms praise God in saying: "You make grass grow for flocks and herds and plants to serve mankind; that they may bring forth food from the earth."[27] In Deuteronomy, God tells His people: "And I will send grass in the fields for thy cattle that thou mayest eat and be full."[28]

In 1872, Senator John James Ingalls of Kansas spoke eloquently on the topic in a speech called "In Praise of Bluegrass," in which he echoed biblical themes. "The primary form of food is grass," said Ingalls. "Grass feeds the ox: the ox nourishes man: man dies and goes to grass again; and so the tide of life, with everlasting repetition, in continuous circles, moves endlessly on and upward, and in more senses than one, all flesh is grass."[29]

To fully appreciate the importance of grass, we must go back much further in time, even before the appearance of *Homo sapiens*. Prehistoric earth was not, as is commonly believed, thickly carpeted from ocean to ocean with damp ferns and towering trees. Ancient flora was highly varied, and from at least 65 million years ago, it included grasses.[30] Probably not coincidentally, and likely due to climatic changes, grasses emerged around the time of the disappearance of dinosaurs, which had previously populated the earth for 160 million years.

From approximately 20 to 10 million years ago, grass began to proliferate in earnest. Changes in the earth's climate are again likely responsible. Initially, carbon dioxide levels dropped, followed by a period of time in which atmospheric carbon dioxide sharply increased, to about 400 ppm (similar to today's level).[31] Amid such fluctuations, grasses had a photosynthetic advantage over other plants. Around the same era, forest-clearing wildfires became common occurrences on the earth. Into these newly opened spaces, grasses (along with forbs, other non-woody vegetation) filled the voids. Resulting from some combination of atmospheric changes and forest fires (scholars debate which factor was more significant), grasses carpeted the globe.[32]

The millennia that followed saw the co-evolution of grasses with a dazzling array of prehistoric animals. These included grazing herbivores and the predators that pursued them. Over much of the globe, including the land that is now the United States, for some five million years roamed enormous populations of "megafauna," great beasts that lived either entirely or mostly from grasses or ate animals that did.

As I gaze out at the open meadows of our California ranch, I sometimes picture the mind-bending menagerie that covered these lands in prehistoric days. These early animals included the western camel, with limbs 25 percent larger than modern dromedaries; the large-headed llama; elk, deer, and antelope; two varieties of bison (including the giant bison, the largest one that ever lived); the shrub ox and woodland musk ox; the mountain goat; the bighorn sheep; two types of horses; and two varieties of mammoths.[33] Mammoths abetted forest clearing both by felling trees on which they were dining and by effectively crushing aspiring saplings with their weighty footsteps. Predators feasting on such prehistoric grazing herds included a lion larger than those of today's Serengeti; a cheetah; a jaguar; a sabrecat (the size of a female African lion); a puma; wolves; and three types of bears.[34]

Like these prehistoric creatures, early humans depended on and co-evolved with grass. By the time *Homo sapiens* emerged, some 200,000 years ago, grasses flourished in nearly every geography and every climate. Some directly nourished humans with their seeds. More important, grasses fed the prey of the human hunter and enabled successful hunting; in open, grassy areas, people could spot, track for long distances, and ultimately kill foraging herbivores.

There is substantial scientific evidence both that humans influenced the way grasses evolved and that grasslands were a major factor in the way humans evolved. *Born to Run*, Christopher McDougall's fascinating best-selling book about traditional people who engage in long-distance running, expounds on the idea, known as the Running Man theory. It argues that it is the human's capacity to run extremely long distances (not just 26 miles, but 50, 75, and more)—farther than nearly any other animal—that was the secret to our ancestors' hunting success. Not speed but endurance was the species' trump card. However, hunting based on outlasting your prey only works if there are vast open ranges where the prey can be spotted and followed even after a burst of sprinting to escape. Thus, the theory argues, open, grassy areas were essential to humanity's success as a species.

Grasslands also provided many other benefits to early humans—ranging from the purely aesthetic to vital self-defense. These included flowers for decoration, fiber for basketry, and creating the unobstructed views that permitted humans living in settlements to see approaching danger from great distances.

These life-sustaining functions motivated humans throughout the world to both create and maintain grasslands. Even well before settling and farming the Fertile Crescent some 12,000 years ago, humans had been aggressively managing their surroundings to support grass. Like the peoples in the Americas discussed earlier, they employed burning and clearing.

As for the world's prehistoric megabeasts, their populations rather suddenly diminished around the same time as the dawn of agriculture, affecting the shape of the world's plants, animals, and ecosystems ever since. At Oxford University in March 2014, scientists from various disciplines focused on the unique role these large beasts had in and on ecosystems. "[E]cologically speaking," the conference's co-organizer said, "it was only a blink of an eye ago that there were megafauna everywhere." He reported that among the scientists in attendance, there was a strong consensus that "megafaunal extinctions did have a huge impact on the structure of ecosystems, which still ramifies through to the ecosystems today." He added: "What we think [of] as natural now are still carrying many disequilibria—or ghosts—resulting from the loss of the megafauna in terms of their structure, functioning and nutrient recycling." Among other functions, researchers agreed that these large animals would have distributed "vital nutrients for plants via their dung and bodies."[35]

Moreover, the loss of the large herbivore and predatory beasts radically altered the evolution of plant and animal populations. So much so that a group of credible U.S. scientists, including some at Cornell University, is actually advocating reintroducing large predators to the U.S. plains. "Their disappearance left glaring gaps in the complex web of interactions, upon which a healthy ecosystem depends," they have stated.[36]

This is part of a whole ecological movement called "rewilding" that has recently emerged, especially in Europe, based largely on the premise that ecosystems cannot properly function without large herbivores. A June 2014 article in The New York Times describes the movement as advocating for the injection of wildlife, "particularly of large herbivores"—without which "untended farmland will become overgrown with thick vegetation that will end up killing off what biodiversity still exists today." The article quotes the communications director for Rewilding Europe as saying: "We need to bring in a few of the parts that are missing because they just aren't there anymore, and that missing part is often the large herbivores."[37] Among projects already under way are reintroduction of the European bison in an area of abandoned farmland in the Romanian Carpathian Mountains and reintroduction of the ibex along a stretch of the Adriatic coast of Croatia where there are two national parks.[38]

By about 100,000 years ago, many of the large beasts had gone extinct, probably mostly from human hunting (although scientists—including those who gathered at the Oxford megafauna conference—are actively debating the impact of the various factors at work). In California, grassland experts believe that as recently as 6,000 years ago, 19 species of large browsing and grazing animals still populated the open lands.[39] Herbivores ate down vegetation while they, along with their predators, trampled the earth with their paws and hooves. Fires, both natural and human-generated, and the presence of these animals, created and maintained California's grasslands.[40]

Cattle, of course, are domesticated animals. They belong to the zoological order Artiodactyla (Greek for "even-toed"), suborder Ruminata (cud-chewing animals that are strictly herbivorous). Most modern cattle are believed to be descendants of a much larger, now extinct wild beast called the aurochs. The book *book* *Animals That Changed the World: The Story of Domestication of Wild Animals* calls cattle domestication "the most important step ever taken by man in the exploitation of the animal world," but notes that how and when it happened are still largely unsettled. Recent research suggests that bovine domestication may have begun as far back as 11,000 years ago, independently in two or more locations, probably India and the ancient Near East.[41]

Cattle first arrived in the Americas in 1493, with Columbus's second voyage, which brought Spanish breeds, mostly as draft animals. Not until 1591 did the original seed stock for Latin America's cattle herds arrive, when a Spanish merchant named Gregorio de Villalobos shipped a small group of Andalusian breed cattle to the New World. In the early 1600s, European settlers of

North America began landing with early British and Continental breeds, primarily for labor and milk.

As settlers pushed south then pioneers headed west, cattle were part of the migration, proliferating particularly in areas with abundant native grasslands. Many of the earliest cattle ranches were in the South, especially Georgia and the Carolinas. "The long grazing season, mild winters, sparsely wooded uplands were especially favorable for beef production," notes a cattle history. "It was said that a steer could be raised as cheaply as a hen."[42]

By the mid-1800s, many cattle ranches were being established on the grasslands of the West. But the bulk of the nation's cattle raising was taking place in the heart of the country. "Prior to the Civil War, the Ohio and upper Mississippi Valley states constituted the center of the beef cattle industry. On practically every farm of this area was a herd of beef [mother] cows."[43]

Europeans spreading across North America encountered grass prairies that had produced the most fertile and abundant topsoil on the planet. Settlers tore open these prairies with their plows to plant row crops. Cutting up the natural sod was tantamount to exploding a vault with our nation's most precious jewels. An old saying goes: "Breaking the sod will make a man, re-making it will break him." Which means, of course, that there is enormous wealth caught up in the grassland soils, but it is one that will be quickly spent when plowed, and is virtually irretrievable.

Poor farming practices eroded American topsoil at an alarming rate. Farmers first plowing up Midwestern prairies had been greeted by chest-high grasses and topsoils 6 feet deep. In 1934, precipitated by years of serious drought and crop failure, a 100-million-acre dead zone on those lands caused epic dust storms as far away as New York City. The center of the country became known as the Dust Bowl. Under Franklin Delano Roosevelt's leadership, the U.S. government created the Soil Conservation Service in 1935 in the hope of reversing the perilous trend.

The new federal agency's field trials soon reinforced history's lessons about the dangers of the plow. Breaking sod and stripping away the earth's protective vegetative cover disrupts soil's life-sustaining systems, depleting its natural fertility, destroying much of its ability to retain moisture, and killing beneficial organisms. The agency's studies showed that the resulting brittle, lifeless soil was becoming less and less productive over time, and suffering severely from wind and water erosion.

Mercifully, by the mid-20th century, Soil Conservation research was revealing solutions to the problem of agriculture. Most important, it showed that regularly rotating croplands into pastures—especially of mixed grasses and clover—reverses much of the damage done by row crop cultivation. Thus, it demonstrated that such meadows not only fed grazing animals, but were also fundamental to creating a permanent farming system that could produce food sustainably over the generations.

Cover crops were already known to help keep soils in place. Dense cover of aboveground leaves of legumes and grasses shield soil from the erosive forces of raindrops and winds, while tangled, fibrous networks of roots beneath the surface hold soil in place, feed soil organisms, and provide sponge-like channels for water.

Legumes like clover take the air's nitrogen and, with assistance from belowground microorganisms, replenish nitrogen to the soils. Most nitrogen enters soil from the atmosphere in a process known as biological fixation, explains Professor David Montgomery. We've already seen the importance of soil's microorganisms in glomalin production. Similarly, bacteria living symbiotically with plant hosts, like clover, reduce inert atmospheric nitrogen to biologically active ammonia in root nodules. "Once incorporated into soil organic matter, nitrogen can circulate from decaying things back into plants as soil microflora secrete enzymes that break down large organic polymers into soluble forms, such as amino acids, that plants can take up and reuse."[44]

Soil Conservation research station investigations showed that mixes of grasses and legumes restored soil's fertility, organic matter, and tilth (soil's texture and ability to retain water), along with controlling insect pests and weeds. Their field trials found grass cover to be 200 to 2,000 times more effective than cultivated row crops in protecting earth from soil losses.[45]

Such research fostered optimism that U.S. farming was on the dawn of a new era of sustainability. USDA's official *Yearbook of Agriculture* for 1948 hopefully described a newly emerging "grassland philosophy," touting the plant's forthcoming Golden Age, and was titled simply *Grass*.

The volume's foreword, by then secretary of agriculture Clinton P. Anderson, calls grass "our alliance with nature" and the "foundation of security in agriculture." He notes that farming based on grass would improve national health and is "a tool against floods and a guardian of the water supplies." The agriculture secretary continued:

> [M]any of the people with whom I have talked look upon grassland as the foundation of security in agriculture. They believe in grass, and so do I, in the way we believe in the practice of conservation, or in good farming, or prosperity, or co-operation. For grass is all those things; it is not just a crop. Grassland agriculture is a good way to farm and to live, the best way I know of to use and improve soil, the very thing on which our life and civilization rest.[46]

Around the same time, the American Society of Agronomy's former president H. D. Hughes authored a hefty tome called *Forages*, in which he wrote: "That adapted grasses and legumes are the chief tools in soil building, improvement and conservation is now generally recognized."[47]

Looking at American food production today, these statements are nothing short of astonishing. They are remarkable both for their prescience of modern soil science and for their misplaced optimism. Historical events soon led American agriculture to veer sharply away from grass. At the close of World War II, munitions plants were converted to manufacturing agricultural chemicals. Use of manmade fertilizers in the United States quickly doubled. Government policy subsidized and encouraged maximum grain output. At the same time, its pro-production policies strongly discouraged permanent pastures, rotating crop fields into grass, diversified farming with animals, and the use of grass buffers.

Simultaneously, agriculture was becoming increasingly specialized and segmented. The discovery of laboratory-produced vitamin D and antibiotics were making it feasible, for the first time, to restrict animals indoors round the clock. Grass habitat for farm animals was soon nearly abandoned in favor of crowded confinement systems, fencerow-to-fencerow plowing, and chemically based fertility, pest control, and weed suppression.

Turning away from grass-centered farming has not only robbed billions of animals of decent lives, but also had severe ecological consequences. Soils and waters have lost their stalwart guardian as millions of acres of grasslands have been plowed, and grass buffers and grass crop rotations all but abandoned. Today, agriculture is the nation's leading source of water pollution and is a significant air polluter, as well, on top of its contributions to global warming.

Especially alarming is the prospect that current agricultural practices, by failing to safeguard our soils, jeopardize our future capacity to feed ourselves. Food production depends on soils created by geological actions over millions of years. But American soils are being blown and washed away at rates far faster than the earth can regenerate them. Scientists estimate that farming that leaves bare soils exposed for much of the year has caused the United States to lose 30 percent of its topsoil over the past two centuries.[48] Other than overpopulation, soil erosion is the single greatest threat to humans' continued ability to produce sufficient food, according to Cornell University's David Pimentel.[49]

Soil erosion rates in the United States are actually "many times higher than official estimates," according to a 2011 report by the nonprofit Environmental Working Group.[50] "Across wide swaths of Iowa and other Corn Belt states, the rich dark soil that made this region the nation's breadbasket is being swept away." Additionally, the report notes that federal policies create perverse incentives that drive farmers away from conservation and toward plowing up their lands. These include the 2007 energy bill mandates for corn ethanol, and the abandonment of landmark soil conservation practices adopted in 1985.

Alarmingly, Americans are still shredding and stripping native grasslands with the plow. Research published in February 2013 in the *Proceedings of the National Academy of Sciences* by scientists at South Dakota State University found that between 2006 and 2011, farmers in the Dakotas, Minnesota, Nebraska, and Iowa had plowed up 1.3 million acres of native grassland in order to plant corn and soybeans. Using government satellite imagery, the researchers concluded that the region's land-use changes—up to 5.4 percent annually—parallel some of the worst deforestations taking place in the developing world. "Between 2006 and 2011, over a million acres of native prairie were plowed up in the so-called Western Corn Belt to make way for these two crops, the most rapid loss of grasslands since we started using tractors to bust sod on the Great Plains in the 1920s."[51] Much of this cropland expansion has been in response to federal incentives for crop-based "bio-fuels."[52] The Nature Conservancy has called grasslands the world's most imperiled ecosystem, noting that (among other things) their demise has serious consequences for climate change.[53]

California grasslands are no exception. In 2011, I attended a gathering at Stanford University of biologists, rangeland ecologists, ranchers, and policy advocates. We were meeting to discuss the ecological value of rangelands and the greatest threats to their continued existence. Several presentations graphically illustrated ongoing conversions throughout the state of rangelands to housing developments and vineyards. Lately, almond groves, which are an especially thirsty crop, seem to be preferred. A recent land deal by Trinitas Partners, a Silicon Valley-based private equity firm, will plow 6,500 acres of "rugged eastern Stanislaus County land from grazing to almonds," according to the *San Jose Mercury News*.[54]

Even putting aside water use and soil erosion, current agricultural methods will no longer be viable in the not-too-distant future. The manufacture of fertilizers, pesticides, and herbicides is fossil-fuel-intensive. Confinement animal operations and feedlots are particularly dependent on cheap, abundant fossil fuels because of their mechanization, and their use of row crop feeds, which are energy-intensive to produce and often transported great distances.

Fossil fuels are a finite, diminishing resource that many experts believe are already more than half expended. More urgently, as PhD soil scientist and dairy farmer Francis Thicke argues, the extreme oil price spikes predicted for the coming decade will render fossil-fuel-based agriculture impractical long before the world's oil runs dry.[55]

Thicke's assessment is shared by soil scientists Lal and Stewart, whose report states: "Intensive agriculture has always been dependent on the energy market because of the energy requirements to manufacture and deliver fertilizer to the farmstead and to operate labor-saving farm equipment." They note that the world experienced a foretaste of the problems to come in 2008 when crude oil prices tripled to about $148 per barrel in a six-month time span.[56]

Food production that can truly be sustained over time will take new policies and approaches. Tomorrow's farming cannot look like the agriculture of today. And transformation is urgent. Rather than haphazardly reacting to future environmental and oil crises, the United States should be engaging *now* in an orderly and deliberate transition away from fossil fuels and toward grass.

To begin the shift, Congress should act forthwith to reinvigorate the 1985 soil conservation standards. It must also forge future farm bills that tie incentives to safeguarding natural resources. As the Environmental Working Group's "Losing Ground" report puts it, we should be "requiring farmers to protect soil and water in return for the billions in . . . subsidies that taxpayers put up each year." This means, among other things, grass rotations for croplands, use of cover crops, and reestablishment of grass buffers. It means fully funding the Conservation Reserve Program (CRP), which provides incentive for maintaining grasslands rather than converting them to cropland.

Agricultural subsidies and incentives should, across the board, foster the use of grass. Farmers should be rewarded for converting croplands to grasslands, rather than the other way around, and for using grass cover crops instead of chemicals, and establishing and maintaining grass buffers. All farm animals, but especially cattle and other grazing animals, should be moved out of buildings and feedlots and reared on grass instead, on meadows, rangelands, or as part of mixed-crop rotations.

Fertility and pest control from grass rather than chemicals comes from biological processes. Using manmade fertilizers, insecticides, and fungicides is food production based on external-ized costs (also known as throwing the garbage over the fence). We need, instead, food production with a full accounting. We need a food system that truly regenerates and preserves resources. With grass we can regenerate fertility and keep away pests without chemicals, breaking away from polluting, soil-eroding systems. With animals on grass, we can reestablish ecosystems closer to the way they evolved and were intended to function. At the same time, we can rebuild soils and sequester carbon. In short, we can raise animals and grow food in ways that sustain natural resources for future generations.

Making grass, rather than chemicals and mechanization, the foundation of our food system is a massive but necessary shift. Farming based on grass would likely involve the United States raising fewer farm animals and entail additional labor (for more animal husbandry and for the planting of cover crops), which means higher production costs. In the short term, it would also mean lost farm revenue in refraining from cultivating every avail-able acre. (It is unhelpful that large swaths of cropland are now rented rather than owned, which disconnects farmers' economic interests from the benefits of long-term conservation.)

The move beyond fossil-fuel-based farming is at once unavoid-able and essential. And with such a transition, cattle, as maintainers of grasslands and converters of grass to meat and milk, will play a central role. Perhaps we will yet fully realize the stalled dream of an American farming founded on grass.

CHAPTER 3

WATER

It's often said future wars will be fought over water, not oil. Whether or not this prediction turns out correct, few question water's importance to human survival. The human body is 60 percent water; we can last just three days without it. Food production is likewise entirely dependent on hydration, often irrigation. A truly sustainable food system will keep water pristine and plentiful.

Yet it is widely recognized that modern agriculture is a poor guardian of our waters. Every year, new books, articles, and studies document the ways agriculture wastes and contaminates water. The federal Environmental Protection Agency has declared that agriculture is the United States' single largest water polluter, blaming agriculture's sediments, nutrients, pathogens, and chemicals for 60 percent of impaired river miles and half of all polluted lake acreage. It is now well established that U.S. farm runoff, primarily from crop production in the Mississippi River watershed, is responsible for a dead zone the size of New Jersey in the Gulf of Mexico.[1]

Beef production is sometimes singled out as a polluter or, more frequently, as a heavy user of water. On countless occasions over the years I've heard or read that it takes a gargantuan volume of water to produce a single pound of beef. Frequently the statement is followed by the conclusion that in an ever-thirstier world, there is no longer room for cattle. I am sure this was among the reasons I stopped eating beef myself. Yet we've already seen how the impacts of cattle on soils and climate vary dramatically depending on how they are raised. Well-managed cattle can be an ecological benefactor, which is no less true when it comes to water.

As an environmental lawyer specializing in water quality, such issues came into sharp focus for me. After graduating from the University of Michigan Law School, I began my legal career as an assistant district attorney in North Carolina, then returned to Michigan and worked in private law practice. In my sixth year as a lawyer, I took a job at the National Wildlife Federation, in Ann Arbor, Michigan, an office focused on protecting the Great Lakes. My time was mostly spent advocating for strengthening regulations and enforcing laws relating to water quality.

In 2000, Robert F. Kennedy Jr. offered me a job at Waterkeeper in New York. The organization is a network of local groups, each working to protect a particular body of water. When I joined, Bobby had been hearing from many of these "keepers" that agriculture, especially industrialized animal production, was their biggest pollution problem. Algal blooms, fish kills, beach closings, and tainted groundwater caused by spills and leakages from industrial animal facilities were all becoming common occurrences. The Clean Water Act explicitly covers "concentrated animal feeding operations," yet the law was barely enforced. My assignment was to change that.

Bobby asked me to launch a "campaign" (as he put it) to fight pollution coming from concentrated animal agriculture operations. Not convinced I would enjoy spending all of my waking hours focused on manure, I hesitated. Then I traveled to communities in Missouri and North Carolina overrun by industrial hog operations.

I met people who'd lived in the same home their whole lives yet could no longer sit on their front porches or hang their laundry outside due to the stenches. I saw cold, lifeless facilities with metal walls and concrete floors holding thousands of sentient creatures. And I saw, and *smelled*, giant festering ponds of liquefied manure behind every cluster of buildings. Manure was leaking and spilling into local streams and rivers, now choked with algae, and periodically filled with dead and diseased fish. Community members had done everything they knew how: written letters, spoken at public meetings, met with elected officials. But government was doing nothing to help. I soon realized this was a fight I wanted to start.

I was witnessing the result of a radical reshaping of animal farming that occurred mid-20th century. Chickens, turkeys, and pigs had long lived in small flocks and herds that spent their days roaming and foraging outdoors, with the nighttime protection and comfort of bedded barns. Nearly all farms had animals, so they were broadly dispersed around the country. Spurred by government policies, together with certain new drugs and technologies, a handful of motivated entrepreneurs brought farm animals in off the land and into crowded metal buildings, concentrated in locations where land was cheap. Independent ownership of livestock and poultry by millions of farmers and ranchers dissolved as a few dozen agribusiness corporations gained control. Grazing and foraging were replaced by concentrated feeds grown and transported, often from distant venues. North Carolina's confinement swine industry, which became the nation's second largest, was built on daily trainloads of corn from the Midwest. The businessmen pushing hog and poultry farming toward the confinement model were all heavily invested in the feed sector.[2]

Crowding, stress, and lack of sunshine, fresh air, and exercise quickly resulted in skyrocketing rates of illness and death in poultry flocks and swine herds.[3] Soon it became routine to add antibiotics to daily rations, drugs that kept the creatures alive in otherwise unlivable conditions. As an added bonus, antibiotics sped up animal growth. The adoption of these practices was enabled by lack of public awareness and lax government oversight.

My job at Waterkeeper was to engage in full-time combat against pollution from such industrial operations. I traveled around the country, investigating the downstream effects of manure spills and leakages. I met and interviewed experts in water biology, public health, and sociology who were diligently documenting debilitating impacts on people and the environment. I read thousands of articles and hundreds of studies about water and air contamination and human health issues connected with animal confinement operations. We began organizing people living near the facilities, providing them information and connections with fellow activists and experts. Eventually, I became an expert myself.

In my research, I discovered that the profound changes to livestock and poultry farming had been reflected in the 1972 Clean Water Act. Congress understood that industrialized facilities (concentrated animal feeding operations, or CAFOs) posed a new threat to water in ways that grass-based farming never had. Kansas senator Robert Dole made the following statement in the *Congressional Record*:

> *Animal and poultry waste has, until recent years, not been considered a major pollutant. The picture has changed dramatically, however, as the development of intensive livestock and poultry production on feedlots and in modern buildings has created massive concentrations of manure in small areas. The recycling capacity of soil and plant cover has been surpassed . . . The present situation and the outlook for the future developments in livestock and poultry production show that waste management systems are required to prevent waste generated in concentrated production areas from causing serious harm to surface and ground waters.*

Soon the campaign I was leading for Waterkeeper started suing polluting confinement swine operations under the Clean Water Act, along with other environmental and nuisance laws.

My favorite part of the job was visiting good farms and ranches. I felt enormous relief and pleasure in seeing animals leading lives that were normal and natural: frolicking in grassy meadows; jockeying for position at the water trough; lying in the sun. These farmers and ranchers spent much of their time with their animals, outdoors, and understood the local climate and ecology. Cycles of regeneration were readily apparent: Animals grazed fields; their urine and manure went directly onto the ground; feed crops and bedding would later be grown in those same fields. Life was bursting forth everywhere I turned.

The work was simultaneously rewarding and invigorating, yet depressing. Obstacles to reform were huge. Political pow-erhouses—food, chemical, agribusiness, and pharmaceutical industries—were heavily invested in maintaining the status quo. Industrial animal operations were ugly, stinking, uninviting places where neither people nor animals were ever seen out-of-doors. Inside the buildings, workers breathed fouled air, and animals were packed together, manure-stained, and listless. The most striking features were putrefying mountains of poultry litter or lagoons of liquefied porcine manure, 12-foot-high chain-link fences, and signs with red lettering blaring KEEP OUT!

Such industrial animal operations are not only off-putting, they're a threat to clean water, both by what's coming in, and by what's going out. Going out the back end is the manure. Any gardener is familiar with the value of manure to growing things. And on a traditional farm with a reasonably sized flock or herd of healthy animals, manure is, indeed, an irreplaceable, life-sustain-ing asset. As we've already seen, returning manure to the earth keeps soils alive and generative; it improves the ground's ability to hold water and carbon; and it enables cultivation of crops, including animal feeds, year after year.

"Livestock manure, when properly used, is a valuable resource that increases the biomass and biodiversity in agricultural sys-tems," notes the Pimentels' textbook *Food, Energy, and Society*.[4] More specifically, manure application enhances the biodiversity of soil-dwelling organisms. Studies in the former USSR have shown that species diversity of larger soil critters (like insects) increased by 16 percent when organic manure was added to experimental wheat plots. Similarly, organic manure added to grassland plots in Japan caused species diversity of soil organisms to more than double in number, and caused a 10-fold increase in diversity in Hungarian agricultural land.[5] Healthy soil communities mean fertility and carbon sequestration.

"Livestock manure is of tremendous value in holding the soil," notes Wes Jackson, founder of the Kansas-based Land Institute. "Its spongy nature absorbs the blows of rain and the water itself." Jackson cites research in Iowa showing that 16 tons of manure applied to a 9 percent slope reduced the soil loss by over 17 tons.[6]

Yet as Senator Dole suggested decades ago, industrial operations transform manure from ecological benefactor to ecological nuisance, and from economic asset to economic liability. It all goes back to animal concentration, which, in industrial operations, is extreme. While a typical hen flock was once a dozen hens, an industrialized facility will often have over one million birds. Where a hog farm traditionally had 25 pigs, many operations now have tens of thousands. With such large animal populations in such small spaces, keeping the facilities in balance with the surrounding environment is nearly impossible.

The waste stream from industrial operations is tainted, infected, and excessive. Jamming so many animals into buildings ensures rampant disease problems, correspondingly high levels of continual antibiotic dosing, drug-resistant pathogens, and massive waste output. The work of North Carolina State University water scientist Dr. JoAnn Burkholder has shown that typical hog effluent contains hundreds of distinct contaminants, including heavy metals, hormones, pesticides, and pathogens.[7] Wastes are stored in huge stinking mounds (for turkey and chicken operations) and in festering football-field-sized lagoons of liquefied manure (for hogs, dairy, veal, and egg operations). The giant manure stores frequently leak into groundwater and run off into surface water, and constantly give off pollutants to the air. So ubiquitous is lagoon leaching that state laws typically even expressly allow a certain amount. In Iowa, a 7-acre lagoon is allowed to leak up to 16 million liters of untreated waste each year. Nonetheless, a study of the state's lagoons found that more than half of them leak in excess of the legal limit.[8] When manure storage structures are emptied, the animal waste is dumped on nearby lands, often overwhelming the absorption capacity of plants and soils, leading to further water contamination.

The other side of the water pollution equation is from concentrated animal operations' inputs. Continually confining animals necessitates delivering them all of their feed. In the United States today, some 55 percent of grains are destined for such operations.[9] Crop cultivation of corn and soy entails plowing, planting, irrigating (sometimes), harvesting, drying, and transporting. Crop production is energy-intensive, the main cause of soil erosion, and the primary source of agriculture's greenhouse gases. In the United States, "[c]orn production causes more soil erosion than any other crop," according to Professor David Pimentel.[10] Irrigated crops divert streams and suck up groundwater reserves. "In addition," the text states, "corn production uses more herbicides and insecticides than any other crop in the United States, thereby causing more water pollution than any other crop produced and therefore is a major contributor to ground water and river water pollution."[11] Corn farming is also by far the largest fertilizer user, with 97 percent of corn acreage grown with chemical fertilizer,[12] according to recent government data. The Gulf of Mexico Dead Zone starkly illustrates the enormous water pollution problem caused by American crop production, especially corn and soy.

In 2011, USDA scientists demonstrated that grass-based dairies are the most ecological.[13] Using extensive field data, the researchers created detailed models for four types of dairies, ranging from total confinement to fully pasture-based. For each dairy type, the model generated estimates all major environmental variables, including air pollution, soil erosion, and water contamination. Additionally, it created estimates for emissions of the major greenhouse gases—carbon dioxide, methane, and nitrous oxide—from both primary production and secondary production of pesticides.

The models showed grass-based dairy farms to be better for the environment in every regard, with the greatest benefits being protection of soils and water. When high-producing dairy cows were kept in barns year-round, the associated sediment erosion from growing corn and alfalfa for feed averaged 2,500 pounds per acre. But with cows foraging on perennial grasslands, their diets supplemented as needed with purchased feeds, sediment erosion dropped 87 percent to an average of 330 pounds per acre. Runoff of phosphorus, a major water pollutant, dropped 25 percent.

Not surprisingly, carbon sequestration was also far better in the grass-based farm. "When fields formerly used for feed crops were converted to perennial grasslands for grazing, carbon sequestration levels climbed from zero to as high as 3,400 pounds per acre every year." The researchers concluded that cropland transitioned to pasture "can build up lots of carbon in the soil and substantially reduce your carbon footprint for 20 to 30 years."

Even without considering the carbon sequestration, total emissions for the greenhouse gases methane, nitrous oxide, and carbon dioxide were 8 percent lower in the grass system than in the confinement system. Ammonia emissions were lower by about 30 percent. The researchers also found that keeping cows outdoors helped reduce fuel use and the resulting carbon dioxide emissions from farm equipment, because producers didn't need to plant and harvest as much feed for their livestock. "Average net farm greenhouse gas emissions dropped about 10 percent by keeping the herd outdoors year-round."

Calculating the overall carbon footprint for every pound of milk produced in each of the four systems, they found that a well-managed dairy herd kept outdoors year-round left a carbon footprint 6 percent smaller than that of a typical confinement herd. Note that this was despite the fact that the confinement operation's cows produced milk at higher rates while the grass operation had smaller cows with lower levels of milk production. "Although the confined cow produced 22,000 pounds of milk every year and the foraging cow produced only 13,000 pounds, the total amount of milk protein and fat produced on the two farms was essentially the same, because the foraging cows produced milk with more fat and protein. In addition, the same amount of land supported a larger number of the small-framed Holstein/Jersey crossbred cows."

The shift toward animal confinement systems has gone hand in hand with the replacement of humans by machines and technologies. From 1945 to 1994, as farm and ranch labor were displaced, farming's total energy use increased by more than 400 percent.[14] Fossil fuels go into agriculture directly and indirectly. About 60 percent of agriculture's energy goes to powering mechanized equipment and vehicles.[15] In feed production, this means machines for plowing, planting, fertilizing, pesticide application, harvesting, and drying crops. In animal agriculture, it also includes automated waterers and feeders, waste flushing and ventilation systems, and manure spreaders.

The other 40 percent of energy in U.S. agriculture is used indirectly, as both the feedstock and the fuel to manufacture agricultural chemicals.[16] Manmade fertilizers had been invented back in 1830, but were barely used for a century. At the end of World War II, munitions factories were converted to chemical fertilizer plants. Between 1939 and 1945, U.S. fertilizer use doubled. "Since about 1950 when the availability of fossil energy became readily available, especially in developed nations, this supported the 20- to 50-fold increase [worldwide] in the use of fertilizers, pesticides, and irrigation."[17]

In 2004, U.S. farmers put 23 million tons of chemical fertilizers and 1.1 billion pounds of herbicides, insecticides, and fungicides on U.S. soils. The most recent Census of Agriculture (published in May 2014) shows that U.S. agricultural operations used commercial fertilizers on 247 million acres; chemical herbicides on 285 million acres; chemical insecticides on 100 million acres; chemicals against crop disease on 35 million acres; and chemicals against nematodes on 14 million acres. As mentioned above, corn farming is by far the largest user of agricultural chemicals among crops in the United States.[18]

The problems with agricultural chemicals are many, with contamination of groundwater and streams at the top of the list. The most comprehensive analysis of water quality in the nation's streams and surface waters was a decade-long study conducted by the U.S. Geological Survey from 1992 to 2001. The startling results show that "at least one pesticide was detected in water from *all* streams studied."[19] In addition, although banned in 1992, DDT (along with other persistent pesticides) was found in fish and bed-sediment samples from *most* streams and in the flesh of more than half of the fish. "Most of the organochlorine pesticides have not been used in the United States since before [this study] began, but their continued presence demonstrates their persistence in the environment," the USGS reported.[20] In groundwater, USGS researchers found pesticides in both agricultural and urban areas. In shallower wells, "more than 50 percent of wells contained one or more pesticide compounds."[21]

Less well understood are the ripple effects of agricultural chemicals on soils. But it's certain they disrupt the web of complex transactions. Previously, I noted that the research of USDA soil scientist Dr. Kristine Nichols has demonstrated that commercial phosphate fertilizer harms glomalin production. Dr. Nichols also advises that every application of chemical insecticides and fungicides indiscriminately kills beneficial soil organisms.[22] We've seen previously that soil's microorganisms are essential to the functioning of soils, including water holding, fertility, and carbon sequestration. There can be little doubt that if we want soils to functionally optimally, the use of agricultural chemicals should be minimized.

Along with agricultural chemicals, industrial animal production has also resulted in huge amounts of water contamination by pharmaceuticals. In 2011, the U.S. Food and Drug Administration reported that 80 percent of all antibiotics used in the United States every year are going to America's livestock and poultry. Of that 80 percent, FDA said, over 90 percent is given to animals that are not sick. Instead, the drugs are added to feed and water to lower feed costs (since antibiotics hasten animal growth) and to stave off diseases in crowded conditions. Up to 75 percent of antibiotics fed to animals will pass unchanged into animal feces and urine, entering the environment in full force and effect.[23] This raises the specter of widespread effects on wildlife and humans that come in contact with the drugs in the environment and on meat. It is well established that such antibiotic use is contributing to the worldwide rise of antibiotic resistance.[24]

The USGS's study on surface waters found widespread contamination of waters by antibiotics. The report noted that "unlike pesticides, which are intentionally released in measured applications, . . . pharmaceutical residues pass unmeasured through wastewater treatment facilities that have not been designed to deal with them."[25]

For pigs, chickens, and turkeys, some of these various types of water pollution can be lessened. Well-run operations, where animals are provided healthy living environments, for instance, do not need to feed antibiotics. But because these animals are omnivores that don't have the ability to survive on grass alone, water pollution related to their feed cannot be eliminated entirely. Grazing, fresh air, and exercise benefit pigs and poultry immeasurably, but even on the best pasture, they require provision of some amount of feed. The most sustainable farms raise omnivorous animals as integrated parts of diverse operations. Ideally, pigs and poultry spend their days on grass, foraging and grazing, and are given some feeds and crop residues, along with surplus and leftover human foods. Such an approach will lead to reduced pollution from feed production, prevent antibiotic overuse, and allow animals to live healthier, better lives.

For cattle, on the other hand, there is a more pregnant opportunity. Feed production—with all its attendant problems of fossil fuel consumption, soil erosion, greenhouse gases, and chemical pollution—can be avoided altogether. We've already seen that cattle have the special capacity to live on a simple diet of grasses and forbs. We've seen, too, that no land provides better protection to water and soils than one densely covered with grass, and that grazing cattle maintain the plants and soils of grass ecosystems. Equally wonderfully, they nourish themselves from their own foraging. The bovine, in other words, requires neither resources expended to *grow* feeds nor the resources required to *harvest* them, nor any of the various forms of pollution that result from those endeavors. This unique attribute makes cattle vital to our global food supply, especially one designed for long-term sustainability. The tragedy of industrialized agriculture is that it has robbed cattle of this, their greatest contribution to our food system.

Confinement mega-dairies, then, must be regarded as the most offensive farming systems. Like industrial swine operations, they deprive animals of exercise and grazing, and they channel enormous volumes of manure into festering storage ponds that leach to groundwater, can spill to rivers and streams, and emit ungodly odors and pollutants to the air. Their soy- and grain-based feeding regimens raise the same pollution and resource concerns as pig and poultry facilities. But worst of all, this bundle of problems is being foisted onto the backs of grazing animals—creatures that indisputably belong on grass.

Like poultry and pig farming, the American dairy has radically changed over the past century, going from small and grass-based to huge and feed-based. In the 1930s, dairy cows were still present on 70 percent of farms, and dairy herds were small. In 1945, the milking herds of the leading dairy state, Wisconsin, averaged just 15 cows. Even by 2002, the state's average was only 71 cows. Meanwhile, dairy herds everywhere were being bred for ever-greater volumes of milk production and were moved off lush pastures to drier western states where cows are continually confined, or may have access only to a small dirt or concrete lot. In 1994, California became the nation's leading dairy state with most milk coming from such large confinement operations. By 2006, California had more than 1,000 dairies with more than 500 cows. Today, it is common to have more than 10,000 cows.[26]

The dairy sector's rupture with grass can be directly linked to mid-20th-century historical events. "[I]n the decades following World War II farmers found that the use of relatively cheap energy, fertilizers, and pesticides, and greatly improved mechanization could improve farm profits," notes an article in *Journal of Dairy Science*.[27] "These inputs, which allowed greatly increased production per cow, were substituted for pasture in the production process," states a report on dairy farming history by Pennsylvania State University researchers. "This trend gave rise to the predominance of confinement dairy practices on farms throughout the United States."[28]

Confinement dairies are inhospitable facilities incapable of providing dairy cows with good lives. Concrete floors and restrictive physical space have led to an epidemic of lameness (35 to 56 percent)[29] among U.S. dairy herds. Meanwhile, the very existence of such enormous concentrations of animals, and their liquefied manure, poses a serious threat to groundwater and surface waters. To give just one example, in 1998, in Washington State, the manure lagoons of two confinement dairies collapsed within a week of each other. Each dumped the contents of an entire lagoon, one spilling 1.3 million gallons of liquefied manure, the other spilling 700,000 gallons, and both spills ending up in the Yakima River.[30] Citizens in the Yakima Valley, an area known for its apple and cherry orchards, became highly organized and skilled at documenting odors and water pollution from confinement dairies. Later, they called on Waterkeeper for assistance.

Working full-time on problems like these from industrial animal operations invariably led me to question the attitude I'd adopted years earlier toward cattle ranching and beef. The more I saw and learned about the way dairy cows, pigs, chickens, turkeys, and egg-laying hens were raised, the more I regarded beef cattle as having the best lives of all animals in agriculture. My friend Paul Shapiro, a senior vice president at the Humane Society of the United States, has famously said that if your eggs come from a conventional, industrial operation, "It's better to eat a steak than an omelet." And increasingly, I was seeing why. The more familiar I became with modern agriculture, the more I began to view it as ironic that the first meat I'd given up in college was beef.

Unlike other animals raised for meat, milk, or eggs, beef cattle start their lives with their mothers, nourished only by nursing and grazing. Most will continue to live in herds, on pasture or rangelands, for about the first year of their lives. Regardless of how they are raised after that, they will never be confined to buildings or kept continually on concrete. For those reasons alone I now believe that beef cattle have by far the best lives of any animals in the food system.

A bit later, I'll talk more about overgrazing. For the moment, I will just remind you of the vast grazing herds that once blanketed the globe, and of Allan Savory's words that most of the world's grazing lands are not being overgrazed, they are being improperly grazed and "over-rested."

Other than overgrazing, the serious concerns about beef production—including water use and pollution—generally relate to feedlots. The majority of beef cattle in the United States raised for meat (as opposed to breeding herds) are, for the latter portion of their lives, confined to dirt lots, commonly called feedlots. Cattle feedlots raise similar concerns as other forms of industrial production: too many animals, too concentrated, leading to animal welfare concerns, more disease spread, odor, and too much manure to stay in ecological balance.

Cattle in feedlots, like all cattle, pee and poop where they are standing. Pens are periodically scraped, manure is gathered and stored, usually in piles. Periodically, it is transported to fields for land application. Some components of their urine will end up in that manure, some in the air. Cattle feedlots, especially large ones, tend to reek to high heaven, and are often the source of huge insect pest problems for neighbors.

Nonetheless, there are important differences from other types of animal facilities. Most important, beef cattle are never crammed into stinking buildings. While a steer's life in a feedlot is undoubtedly boring, and lacks grazing, at least he is outdoors, on soft ground, with room to move about. From an environmental standpoint, it's important that beef cattle feedlots neither liquefy manure nor store it in lagoons. Handled as a solid, manure poses less risk of leakage to groundwaters and spilling to surface waters, and there's less volatilization to the air. That said, the larger the feedlot, the more animals, the more manure, and the greater the likelihood that any of these potential problems will occur.

It's popularly claimed nowadays that beef feedlots and feeding grain to cattle are new ideas. This is simply untrue. Neither ancient European nor early American cattle husbandry was strictly grass-based. According to *A Short History of Farming in Britain,* at least as early as the Middle Ages domesticated cattle were fed exclusively "hay and corn" (*corn* in the British sense meaning "grain") during the winter.[31]

Moreover, in the newly forming United States, cattle typically grazed on grass from spring to fall then ate hay and grains in the weeks or months prior to slaughter. Describing the Ohio and Mississippi Valley regions of the mid-19th century, a history of U.S. beef cattle explains that cattle were kept until three or four years of age, at which point they were put into feedlots and fattened on corn. "Grass was still abundant and relatively cheap and constituted the sole feed during the summer and fall; while hay, either timothy or prairie, supplemented with a liberal allowance of shock corn, formed the winter ration."[32] By the 1870s, U.S. beef cattle breeding herds were being kept in the open ranges of the Far West, whereas cattle fattening (largely on corn) was mostly taking place in midwestern feedlots. While it is frequently suggested nowadays that feeding corn to cattle is a post–World War II invention, in fact nothing could be farther from the truth.

For environmental reasons, and in my gut, I always prefer to see any creature, and especially grazing animals, scampering among rolling, grassy hills. But I am unwilling to uniformly condemn cattle feedlots. I fully appreciate what my good friend Will Harris, a Georgia grass-fed cattle farmer, has colorfully stated about feedlots: "Sending cattle to a feedlot is like raising your daughter to be a princess and then sending her off to the whorehouse." Yet, having visited several well-run smaller-scale facilities, it is clear to me that feedlots are highly variable. When relatively small in size, where only mature cattle (not calves) are accepted, careful feeding regimens are strictly observed, and runoff is carefully controlled, a cattle feedlot can be acceptable both in terms of animal welfare and the environment.

Comparatively little of my time at Waterkeeper was spent dealing with cattle-related problems. The reason is simple: Most beef cattle and many dairy cattle (especially in the Upper Midwest) still reside on grass. And when animals are reared on grass, water pollution is minimal.

Despite the prevalence of feedlots in the United States today (the most recent Census of Agriculture says there are now 13,734, most quite small), it's still true that beef cattle are generally raised similarly to how they've been raised for millennia: foraging their own feeds from grasslands and being fed grain for fattening prior to slaughter.

The breeding animals—mother cows (which are 28.9 million of the 89.9 million cattle in the United States[33]) as well as bulls, most of which are in the western United States—typically spend their entire lives grazing rangelands or pastures. They are often moved, especially in colder climates, to higher elevations for some part of the spring, summer, and fall, then led down to areas near the home ranch, and supplemented with hay, over the winter. Many cattle will graze the stubble of field crops in the fall. Cattle mating and calving happens with little human interference, usually out on the range. Mother cows have a single calf each year and have the pleasure of providing their baby a mother's care. Beef calves have the pleasure of being reared by their mothers, and stay with them, on pasture, for the first several months of their lives. Many calves also stay on grass for some time after weaning, even if they eventually go to feedlots. The mother cows, bulls, calves, and yearlings are the black or reddish-brown animals grazing the hills and fields you see on nearly any American road trip.

The black-and-white spotted cattle you may see grazing are generally dairy heifers, meaning future dairy cows. In some parts of the country, especially in Wisconsin, you will also see mature dairy cows out on grass. Most dairy heifers, though, once they are mature enough to enter the milking herd, are taken off grass and confined to buildings from that point on. There, like all animals raised under the confinement model, they are fed concentrated feeds like soy, oats, and corn, often mixed with some alfalfa hay.

Again putting the overgrazing issue aside for the moment, the net effect of cattle in the food system is a benefit to the world's waters. Previous parts of this book show there is nothing better at absorbing, holding, and filtering water than lands densely covered by grass, and that there is no better way to keep those grasslands healthy than by periodically grazing them. Grass-based farming systems also create filters for water going to rivers, streams, and groundwaters. Research in California's Sierra foothills has shown that "grazing can actually enhance the ability of riparian vegetation to filter nitrate out of surface waters." In field experiments comparing test plots of grazed and ungrazed land in annual grassland-oak savanna, soil water from plots with grazing removed for only two years contained up to five times more nitrate than soil water from plots that were grazed. The reasons for this are not fully understood. Researchers hypothesize the effect may be due to reduced nitrate use by plants not subjected to grazing.[34]

I am not suggesting here that there have been no incidents when cattle on grass have caused water contamination. Of course there have been, plenty. Rather, my point is that well-managed grazing does not cause water pollution and that the *overall effect* of having grazing cattle as part of the food system is a net positive for water in terms of both quantity and quality. Compared with nearly any other land use, and certainly with croplands, grasslands are far better at holding soils, making nutrients available for plants, keeping pathogens, sediments, and nutrients from entering groundwater and surface waters, and filtering rainwater. The Land Stewardship Project of Minnesota even recommends permanent pasture as a way to protect waters, stating: "Permanent pasture for grazing livestock can be an ideal choice for minimizing water pollution."[35] Fundamentally, then, a food system that includes cattle will be a better guardian of waters than one that excludes them, and the more time the cattle spend on grass, the cleaner and more "effective" waters will be.

The other side of the water issue is quantity. In recent years, many people have come to accept the idea that beef is particularly and unacceptably water-intensive. The advocacy material of vegan and environmental groups often places water usage at the very top of the list of indictments against beef. Water usage is generally calculated with simple arithmetic adding together three numbers—a volume of water a steer is said to drink in his lifetime, a volume of water said to be used to grow feeds a steer eats in his lifetime, and water used in slaughter and processing. The figure most commonly cited (also often used in mainstream press) is that it takes 2,500 gallons of water to produce a single pound of beef (which converts to about 20,820 liters per kilogram).

Occasionally you'll encounter even much higher figures, such as 12,009 gallons per pound of beef. The latter number, said to be from a book edited by Cornell's David Pimentel, appears in a chart on a vegan website that also indicates a pound of potatoes takes just 60 gallons of water while wheat takes 108 gallons. "As you can see from Professor Pimentel's figures," the website comments on the chart, "it takes roughly 200 times more water to make a pound of beef than a pound of potatoes."[36] But is the matter really so simple?

Of course the foods selected for this chart are designed to create the impression that beef is an extreme water guzzler. But calculations from other sources (unconnected to the meat industry) show some of the most water-intensive foods eaten by humans include common staples like rice, which takes 3,400 liters of water per kilogram (21 percent of the water used in global crop production), and sugar, a completely unnecessary (and harmful, as I'll talk more about later) food additive, which takes 1,500 liters of water per kilogram.[37]

Putting aside momentarily the vast nutritional differences among the foods on the chart I just mentioned, how believable are these water usage numbers? On its face the chart *seems* legitimate, since it apparently comes from a textbook edited by Pimentel and other university professors. Given that I cite his writings in this book, it's obvious I consider Dr. Pimentel a credible source. Pimentel's figures about energy and water in the food system are often touted in anti-beef writings. Yet, as noted previously, Pimentel is not an advocate for veganism. And he himself, a scientist, openly acknowledges the limitations of such numbers. In the chart's annotations he states that the beef figure was calculated from an animal that consumed 100 kilograms of hay and 4 kilograms of grain per pound of beef produced, and that it assumes a generalization that 1,000 liters of water were used to produce 1 kilogram of hay and grain.[38] (Interestingly, if you do the conversion to metric, the wheat figure on the chart said it took 900 liters of water per kilogram of wheat, but here, when calculating grain for the steer's consumption, the number was elevated to 1,000 liters.) Although this 1,000 liters appears to include rainwater, most people reading it and citing it will take it to mean irrigation water. If you exclude rain water (which I believe you should), Pimentel's data would be wildly off for animals fed hay and grain that were grown *without* irrigation, which is the way the vast majority of U.S. grain and hay is, in fact, grown.

Equally important, Pimentel implicitly recognizes that the figure is inapplicable to cattle that are *not* fed grain and hay. Thus, he acknowledges that the numbers have limited application for cattle raised entirely or mostly without feeds—in other words, cattle raised on non-irrigated pastures and rangelands, a not uncommon scenario in the United States and world today.

To me it is clear the water usage figure needs verification from another credible source. I rarely rely on self-reported data from any industry. As you read this book you will notice that almost none of its information, environmental or health, is taken from the beef industry itself. When it comes to analyzing the water usage of cattle and beef processing, however, the most credible and thorough analyses have been done by agricultural colleges. While implicitly linked with the beef industry (and thus somewhat less objective), they are the entities most intimately familiar with every life stage of the animal and every step of the husbandry and meat processing.

In response to concerns over water conservation, researchers from the University of California–Davis undertook an extremely detailed look at the water that goes into making beef.[39] Whereas the water usage figures you typically see from advocacy sources are based on worst-case scenarios, the figure compiled by UC Davis is surprisingly balanced. Rather than showing the least possible water usage, it strives to achieve a number that actually reflects *typical* water usage for a *typical* piece of beef. In the nine-page analysis, it goes, in excruciating detail, through every step of an animal's life, taking every variable into account. Among the factors considered are where cattle were located, average temperatures, the number of days on feed and feed types, typical grain types, how many acres of each type of grain would have been irrigated, and so on. At the end of it all, the UC Davis researchers concluded that a kilogram of boneless beef requires 3,682 liters of water to produce (or 441 gallons per pound). Interestingly, if you remember the numbers from the environmental report presented above, that's just about the same amount of water it takes to produce a kilogram of rice. Having read UC Davis's full report, I find the number credible.

Moreover, if we specifically consider grass-raised cattle and take nutritional value into account (and for such comparisons to make any sense, it *must* be), it becomes even clearer that beef is not *inherently* more water-intensive than other foods. Author Lierre Keith, who returned to meat consumption after living 20 years on a vegan diet, does a good job breaking down this issue in her book *The Vegetarian Myth*. She writes:

> On pasture, beef cattle will drink eight to fifteen gallons of water a day. The average pasture-raised steer takes 21 months to reach market weight. That's 40,320–75,600 pounds of water total for an entire cow [sic]. That's 450–500 pounds of meat, with another 146 pounds of fat and bone trimmed off, which in an earlier, saner era would have been valued for food as well. Taking the mean of 475 pounds, the midpoint of 57,960 gallons yields a figure of 122 pounds [sic] of water per pound of meat [a figure that excludes the organ meats, the most nutritionally dense part of the animal].[40]

(Note that Keith's calculations show that it takes 122 gallons of water per pound of grass-fed beef. The use of the word "pounds" is a typo in the book.) A bit farther on, Keith adds nutritional value to the equation, looking specifically at the oft-made comparison with wheat. She points out that not only is beef much more nutritionally dense, but its nutrients are also far more usable by the human body (a point I explore more fully in the second half of this book).

> The beef contains almost twice as many calories (592 v. 339, per 100 grams) . . . For wheat, sixty pounds of water produces 1524.45 calories, or 25.7 calories per pound of water. For grassfed beef, it's twenty-two calories per pound of water. And there's more than simple energy: those beef calories contain more nutrients, especially essential nutrients protein and fat. The numbers on those are 21g v. 13.7g and 8.55g v. 1.87g respectively. It's also crucial to understand that the protein in beef contains the full spectrum of necessary amino acids and is easy for humans to assimilate, while the protein in the wheat is both low-quality and largely inaccessible because it comes wrapped in indigestible cellulose.[41]

These figures show how wrong it is to treat all methods of cattle husbandry as the same, and it's certainly wrong to use worst-case scenario figures and pretend they are typical. Depending on where feeds are grown, and the type of feeds used, the water usage numbers can vary dramatically. If you omit rainwater and use not a worst-case scenario but an average, the number plummets. Keith's analysis and the Pimentel figures cited earlier both clearly show that water usage mostly occurs in the growing of feeds. In my view, such water usage calculations make no sense when applied to cattle raised on non-irrigated feeds, let alone cattle raised on forages, without feeds. When differing nutritional qualities of the foods are added into the comparison, it becomes plain that a straight pound-to-pound comparison of beef with wheat or potatoes is like comparing apples and oranges.

Furthermore, there is another crucial aspect of the water usage question to consider: the fate of water *after* it hydrates plants or animals raised for human food. From both a resource usage and a broader ecological perspective, figuring out how much water a plant or animal uses in its growth should be only one part of calculating the impact a particular crop has on water resources.

In every sector, water employed ends up somewhere. In manufacturing, it will likely be discharged via tailpipe into a river or lake. In agriculture, water used to hydrate grains or animals will be released to the environment in diffuse ways. Yet that water, too, through its presence or absence, and whether polluted or pristine, goes on to affect wildlife and groundwater, lakes and streams, and even air.

As I've already shown, water landing on plowed fields will wash away more soil and pick up more contaminants as it flows off fields. Rainwater or irrigation water falling on conventionally grown fruits, grains, and vegetables will be also contaminated by agricultural chemicals, including pesticides, fungicides, and fertilizers. Water that runs off or evaporates provides no further support to the life in the immediate ecosystem. (This is like Savory's "ineffective rain.") On the other hand, water retained by plants and soils continues to provide life-giving hydration to surrounding organisms and thus supports the entire ecosystem. (This is like Savory's "effective rain.") Water that passes through cattle on grass ends up hydrating the grass as well as the subsurface communities of insects and microorganisms after hydrating the animal. In other words, that water goes on to support many other life-forms in the grassland ecosystem. When comparing beef's water usage with other foods, such differences must be taken into consideration.

When we consider the best ways to safeguard water in our food system, it's essential to keep in mind that there is no better way to preserve and to clean water than to have it land on grass-covered earth. Research in the Midwest has shown that compared with cropland, perennial pastures used for grazing can decrease soil erosion by 80 percent and markedly improve water quality.[42] In fact, there is no better way to filter runoff than through healthy grasslands. University of Georgia research shows that runoff rates of croplands are 5 to 10 times higher than those of perennial grasslands; even more dramatically, erosion rates from croplands are 30 to 60 times higher.[43] Grass-covered lands benefit ecosystems by holding much more water than croplands.

Cattle enable grass to be a foundation of our food system. Whether grazing on rangelands, annual or permanent pastures, grass buffers, or cover crops, cattle make it possible to produce food from densely vegetated lands that resist wind and water erosion, effectively retain water, and simultaneously support countless species, both above- and belowground.

In the earlier global warming discussion I noted that, depending on farming methods, climate impact can vary by a factor of 10. In a similar fashion, the quantity of water used to raise a food is highly variable depending on how and where it's grown. Whatever plants or animals are farmed, water can be used wisely or wastefully.

Later, I'll describe in detail how we raise cattle on our own Northern California ranch. Here I'll just provide an overview of our approach and note that we use little water, and use it carefully. Our cattle herd is grazed on sections of land in an area that totals about 1,000 acres. All of this land is categorized as natural grasslands, which, in another era, was maintained by giant herds of wild elk and deer. And long before the elk and deer there was a rich diversity of large beasts that populated this land and maintained broad, open areas of grass.

The climate here is Mediterranean. This means temperatures are moderate year-round and nearly all rainfall comes during cool months—roughly from late October to early April. We have little heat (even in summer), and it is often windy. The topography is rugged—hilly with deep gullies and ridges.

The weather and the topography mean that the acres making up our ranch cannot and should not be used for crop production. Yet they are ideal for grass. Many of the grasses that flourish here are known as cool-weather grasses, because they have the unique property of being able to grow in those cold, rainy months.

Our cattle are never fed grains and live their lives year-round on grass. Naturally occurring vegetation is always their primary source of nourishment. We do no plowing, no planting, no chemical applications to our land, and no irrigation. We withdraw no water from wells or municipal water sources and divert no rivers or streams. The only water that ever hydrates the dense vegetation covering the ground of our ranch is rain falling from the sky.

We gather some of the rainwater in a retention pond. This is then distributed via a gravity system to troughs dispersed around the ranch. The water that goes to our home, and to the animals to drink, is collected rain.

Every time one of our cattle pees or poops, we celebrate. That animal's waste falls directly onto our pastures, returning moisture and decomposed vegetation (as well as nitrogen, phosphorus, and beneficial bacteria) directly to our grasses and soils. This, combined with the pruning and smushing of biological materials into the ground, enables the grasses and other vegetation here to thrive. It is a cyclical, regenerative system that works well not only for the cattle but for wild plants and animals as well.

Every situation where cattle are raised is unique. Our ranch is a living display of how land can be used to keep cattle without intensive water use or water pollution. We are proud of the way we care for our animals and our land, but we do not consider ourselves exceptional. I've been to dozens of other farms and ranches where cattle are being raised well and appropriately and in such a way that they conserve and protect water. In many regions of the United States and around the globe there is nothing more appropriate for the topography and climate, as a livelihood or a food source, than cattle tending.

All food production will have impacts on water resources. Nothing about cattle is inherently more polluting or more consumptive than other forms of agriculture, especially when you take the nutritional value of food produced into account. Well-managed cattle are neither polluting nor overly consumptive of water. When cattle are raised well, and raised on grass, the overall impact of their presence is a benefit to the earth's water resources.

CHAPTER 4

BIODIVERSITY

I'VE OCCASIONALLY HEARD IT SUGGESTED that biodiversity—the degree of variation among an ecosystem's plants and animals—is harmed by the presence of cattle. This discussion will be quite brief because I don't think I need to vigorously refute this particular criticism. Even among people who seem to loathe beef, this notion has fallen out of favor. The FAO's *Livestock's Long Shadow* report, amid a sea of criticism of cattle, states: "There is growing evidence that both cattle ranching and pastoralism can have *positive* impacts on biodiversity."[1] Indeed, there is now abundant scientific research showing that cattle grazing does more to help biodiversity than harm it. Although the public discussion of cattle and beef generally lags a couple of decades behind the research, on this issue it's more in sync.

When considering biodiversity, you can look at everything from microscopic organisms to elephants. Earlier, I showed that the biodiversity of soil organisms is enhanced both by the existence of grass and by grazing. Anecdotally, cattle ranchers using soil-building methods, like those advocated by Allan Savory, are reporting seeing more biodiversity as they improve their soils. "We've got more wildlife than ever and more livestock than ever," says cattle rancher Jerry Doan of North Dakota, attributing this to cover crop mixing, more litter, and biodiversity starting at the soil level.[2]

Cattle keeping can be hospitable to wild plants and animals. In sharp contrast with the plowing of land and cultivation of crops, the types of disturbances caused by cattle are like those the globe's ecosystems have experienced for millions of years. Paul Krausman, University of Montana wildlife conservation professor, has noted that some of the greatest threats to wildlife today are tillage agriculture, urban sprawl, tree and shrub invasion, and energy development, all of which "result in broad-scale loss and degradation of habitat that overwhelms management of remaining fragments."[3] Krausman also points to "a host of vegetation manipulations" that harm wildlife, particularly ground-nesting birds. In addition to agricultural tillage, these manipulations include "herbicide application, mechanical sagebrush removal, and over-prescription of fire in xeric landscapes."[4] It should be noted that these are land management practices often used as alternatives to grazing. "Wildlife managers in the Great Plains readily acknowledge the importance of livestock grazing to conservation," Krausman notes, "because ranchers whose operations remain profitable are less likely to convert native prairie to cropland."[5]

In contrast with those inherently problematic land uses, "a number of studies have recorded higher biodiversity in grazed relative to ungrazed systems."[6] Just as Allan Savory has so well articulated, these studies have shown that grazing is not only a beneficial but in fact a *necessary* "disturbance" in grassland ecosystems.[7] And as Savory also points out, since the world's wild herds have been largely extirpated, domesticated livestock must be their proxy. In the words of Nature Conservancy scientist Jaymee Marty, cattle can now "serve as a functional equivalent to large herbivores that historically grazed grasslands and play an essential role in maintaining biodiversity."[8]

Previously, I mentioned a 2014 University of Maryland field study with 40 test plots that found the least plant diversity occurring on plots where grazing had been excluded. Notably, that study also concluded that the beneficial effects on plant diversity were greatest with grazing by large animals, wild and domesticated: cattle, pronghorn, and elk on North America's Great Plains; wildebeests and impala on Africa's Serengeti Plain.[9]

Several earlier studies had reached similar conclusions. A 2001 study in the journal *Plant Ecology* by Richard H. Hart, a senior range scientist at the USDA's High Plains Grasslands Research Station, describes field tests from several Colorado rangelands with varying amounts of cattle grazing—no grazing, light, moderate, or heavy grazing—each practiced for over 50 years. Hart reports that plant diversity was lowest where cattle had been excluded. He notes, too, that this type of grassland, known as a shortgrass steppe, when moderately or heavily grazed by cattle "was similar to and probably as sustainable as steppe grazed for millennia by bison and other wild ungulates."[10]

A long-term study by researchers at University of Nevada, published in 2004, also found more plant diversity with grazing than in areas where grazing had been excluded.[11] In 1934, when the Taylor Grazing Act was adopted, multiple test sites that entirely excluded grazing were established, 16 of which remained when the study was conducted. To ascertain the effects of grazing of large ungulates, the research team conducted a meticulous examination of the ecologies both inside and outside the defined study areas. The researchers then subjected their findings to a thorough statistical analysis. They concluded that "light-to-moderate grazing in the Great Basin certainly has no ill effects on the ecosystem," and that the grazed areas had a greater variety of plants.[12]

As these field studies suggest, real-world investigations into grazing's impacts have often found the test plots from which cattle were entirely *excluded* the most revealing. A university textbook on grasslands notes: "A growing body of research shows that livestock grazing can enhance biodiversity. To a surprising degree, this research comes from cases in which, as part of conservation efforts, livestock grazing was removed, and subsequently, species or habitats of interest disappeared."[13] The unexpected results—that land or particular species' populations are further degraded without the cattle—should no longer surprise anyone. (Of course, had they asked Allan Savory, he could have told them all of this decades ago.)

A telling U.S. example of such counterintuitive outcomes involves California's vernal pools. These are ephemeral, shallow ponds that appear in flat, grassy areas during the rainy season. Attention was first brought to their ecological importance by a PhD botanist and San Diego high school science teacher, Edith Purer. After years spending her summers surveying them, Purer presented a report on California's vernal pools to the Ecological Society of America in 1937. Only decades later was their uniqueness fully appreciated. In the 1970s, more than 100 vascular plant species and more than 34 crustacean species were cataloged in vernal pools, and it was discovered that half of the life-forms inhabiting them are found nowhere else on earth.[14]

Meanwhile, since some ranchers attempted to fill them, and cattle stepped into them, grazed their grasses, and drank from them, many environmental advocates came to regard cattle ranching as an inherent and serious threat to vernal pools. "A major controversy surrounding the conservation of vernal pools concerns grazing," a Nature Conservancy document noted. "Some argue we need to remove cattle from the vicinity of these threatened habitats, and others argue cattle should be used to manage vernal-pool grasslands. In fact, cattle grazing was implicated as a major contributing factor to the decline of four vernal-pool crustaceans listed under the U.S. Endangered Species Act (USFWS 1994) with little to no supporting scientific data."[15]

The dearth of scientific evidence was finally addressed when Jaymee Marty, lead scientist for The Nature Conservancy's Central Valley and Mountains region, began conducting detailed field tests to evaluate the effects of cattle ranching on the state's vernal pools. Her study involved 24 natural vernal pools, of varied sizes and depth, across two different soil types. On some study sites, grazing was unrestricted, on some it was seasonally restricted, and on some it was completely removed.

The findings of this careful, credible, non-industry-funded study shocked many in the environmental community. Marty reported "significant negative effects on the native plant community, pool hydrology, and the aquatic invertebrate community with the removal of grazing." Correspondingly, she found "that livestock grazing played an important role in maintaining species diversity." The report concluded: "Grazing should be considered one of a variety of important tools for land managers interested in the maintenance of biodiversity."

Interviewed in 2009 about these unexpected (by some) results, Marty, who conducted a 10-year study of cattle grazing and vernal pool grasslands to determine the best approach to conservation and restoration of the pools, said: "I learned that grazing is important to maintaining these unique grasslands. Grazing keeps the non-native grasses from crowding out the native plants and sucking up all the water."[16]

Marty's study write-up also analyzed the most likely explanation for these results. Wildlife conservation professor Krausman notes that large grazers have been essential to the evolution of the globe's ecosystems for millions of years. "Indeed, many, if not most, ecosystems rely on grazing by native ungulates to influence vegetation structure and composition."[17] Because California's ecosystems evolved with large grazing animals, the impacts of cattle on the land, similarly to historic grazing herds, have favorable effects for the ecosystem's functioning. Marty explained:

> *Why is cattle grazing so clearly beneficial to biodiversity in these vernal pools, when conservationists often advocate severe grazing restrictions? The answer to this question comes in part from the fact that California grasslands have a long history of extensive grazing dating back to the Pleistocene but were most recently grazed by herds of tule elk* (Cervus elaphus nannodes) *and pronghorns* (Antilocarpa americana) *before livestock introduction in the late 1800s.*

Hence, the pool species are adapted to some level of grazing. In addition, the plant species composition of California Central Valley grasslands has changed significantly since European settlement and is now dominated by exotic annual grasses. Thus, a long history of grazing coupled with the altered plant community yields a system that is now adapted to the changes brought about by cattle and one that becomes quickly degraded when cattle are removed.[18]

Similar findings have been made by scientists studying various land management approaches in the federal Conservation Reserve Program. The program allows some, highly restricted, grazing on areas taken out of crop production. Natural Resources Conservation Service field scientists have noted that grazing helps other animal and plant species. Kevin Reynolds, district conservationist for the NRCS in Decatur County, Iowa, has seen, year after year, the effects of grazing compared with fallowing without grazing, and has observed the value of grazing for songbird populations, in particular. Grazing, Reynolds says, reduces the thatch that makes it so hard for young birds to get around in older stands that haven't been disturbed. He says that properly timed and controlled grazing leads to better overall wildlife habitat. It also protects water and soils, and controls woody vegetation. "The year after grazing, I've seen a noticeable difference in the number of bugs and in the number of birds that eat them on [this] land," Reynolds says, adding, " I'd like to see more grazing here."[19]

Other studies of CRP land have shown the value to various wildlife species of taking land out of crop production and allowing the growth of grasses. Studies have revealed that grassland birds are declining more than any other bird group in North America. Returning cropland to grass has helped protect and reestablish a number of wildlife species, including the bobwhite quail, swift fox, short-eared owl, Karner blue butterfly, gopher tortoise, Louisiana black bear, eastern collared lizard, Bachman's sparrow, ovenbird, acorn woodpecker, greater sage grouse, and salmon.[20] In a similar vein, Austrian studies have shown that grazing improves habitat for wading birds.[21]

Other research has shown positive impacts from grazing on many rare wildlife species. The ways these species benefit from grazing include alteration to grassland structure to create shorter vegetation, more openings, and more structural heterogeneity in general than is found when livestock are excluded. Specifically studied animal species found to benefit from livestock grazing include burrowing owls, a variety of beetles, kit foxes, kangaroo rats, blunt-nosed leopard lizards and other lizards, ground squirrels, and San Joaquin antelope squirrels.[22]

The very existence of grazing areas supplies essential habitat to wildlife. Pollinators, which are indispensable to both wild ecosystems and the human food system, are a good example. According to the U.S. Fish and Wildlife Service, over 75 percent of flowering plants—including apples, almonds, and squashes—need an animal pollinator.[23] Globally, about three-quarters of crop species depend on animal pollinators to produce fruits or seeds, and over 90 percent of vitamins A and C in foods are derived from pollinator-dependent crops.[24] Yet domesticated bee populations are plummeting, suffering from a phenomenon called colony collapse disorder (CCD). Statistics released in March 2013 show a 40 to 50 percent decline in bee colonies over the preceding year.[25] Studies now strongly support the conclusion that domesticated bees are being killed off by a certain class of pesticides (neonicotinoids), although the issue is still being debated.

Especially in light of collapsing colonies, wild pollinators have enormous significance for our food supply. About one-third of crop pollination is done by wild pollinators.[26] And nothing is more important to their survival than preservation of their rapidly disappearing habitat.

And where do these wild pollinating animals live? Rangelands. The very same rangelands that exist because they are being grazed by cattle. "Preserving rangelands has significant economic value, not only to the ranchers who graze their cattle there, but also to farmers who need the pollinators," biology professor Dr. Claire

Kremen and her fellow UC Berkeley researchers conclude in their study of wild pollinators' significance to the food system.[27] In other words, we need wild pollinators and wild pollinators need cattle. "[T]he use of rangelands for sustainable livestock production has the potential to ensure the maintenance of wildlife habitat, especially when compared to energy development and urbanization, which will ensure that wildlife habitat will persist into the future," notes Professor Krausman.[28]

A 2014 study by researchers from Stanford University urges the design of future food systems with wildlife in mind. Rather than focusing on wildlife preserves, they argue, farms and ranches should become part of the strategy for safeguarding the world's wildlife. "Countryside biogeography is an alternative framework, which recognizes that the fate of the world's wildlife will be decided largely by the hospitality of agricultural or countryside ecosystems."[29]

All human activities have variable effects on other animals and plants, depending on how they are carried out. Of course, none of this research shows that every ranch or farm raising cattle benefits biodiversity. For example, ranchers filling in vernal pools are undoubtedly damaging those ecosystems. But that's an example of poor land stewardship rather than an indication that ranching inherently poses an environmental problem.

Research now shows that having cattle on the land creates and sustains habitats and whole ecosystems, which benefits countless plants and animals. And it suggests that the best methods of cattle ranching can have a significant positive effect on the world's plant and animal diversity. Research like The Nature Conservancy's vernal pool studies are helping to deepen the human understanding of the best ways to manage cattle on the land so they fill the ecological void left by the world's disappeared grazing herds. Herds of cattle—large grazing animals over which humans have substantial control—not only provide invaluable nutrition to humans, but also present us the opportunity to create the highest-functioning ecosystems.

CHAPTER 5

OVERGRAZING

EARLIER I NOTED THAT FOR DECADES "overgrazing" was the most commonly heard criticism of cattle and beef. Implicit in the word were the ideas that livestock grazing is responsible for much of the globe's denuded landscapes and desertification, and that the problem boils down to far too many animals eating too much vegetation and doing too much trampling. Books like Jeremy Rifkin's influential *Beyond Beef* urged people to stop eating beef almost entirely based on the argument that the American West and much of the globe has long suffered injury from serious overgrazing. Rifkin portrays an American cattle rancher who is despoiling the environment, one who is unregulated, government-sanctioned, and feeding at the public trough.

Soil erosion and desertification are very real. Just as I do not doubt climate change, I have no doubt that erosion and desertification are happening and have been largely caused by human activities. Globally, erosion is occurring more than 20 times faster than the geological rate, and in some areas the situation is much worse.[1] "The estimated rate of world soil erosion now exceeds new soil production by as much as 23 billion tons per year." At this pace, if left unabated, the world will literally run out of topsoil in little more than a century.[2] Meanwhile, "The [worldwide] rate of desertification is estimated at 5.8 million hectares (Mha) per year," notes one credible source.[3]

Yet the assessment that these problems result from too much grazing—historically and today—is usually done unscientifically. The real question is not whether these phenomena are occurring, but what are the true causes. Equally important, what can humans do to stop or even reverse the trends as we feed ourselves in the future?

Looking back at the history of areas that were once fertile but are now deserts, it's routinely stated (with scant evidence) that overgrazing was the cause of the transition. Regarding the Middle East, Steven A. Rosen, professor in the archaeological division at Ben-Gurion University, Israel, says, "It is a common conception that desertification and the destruction of previously fertile and productive lands are often the result of pastoral nomad activities, in particular through overgrazing by goat and sheep herds." However, the assumption has not withstood scrutiny. Rosen notes that a critical analysis, using archaeology as the primary tool of historical reconstruction, showed "no evidence for causal links between pastoral nomadism and desertification" and "suggest[s] that declines in pastoral nomadic presence in the desert are related primarily to changes in the sedentary infrastructures with which the nomadic systems interacted."[4]

Grazing, more often than not, seems to get lumped in with damage from other human activities, particularly crop cultivation. I mentioned earlier, for example, how Professor David Montgomery's book *Dirt: The Erosion of Civilizations* documents a worldwide pattern: poorly managed crop cultivation leading to soil exhaustion and erosion followed by grazing animals put on the land. Livestock are then (incorrectly) blamed for the land's poor condition. The book chronicles observations of 19th-century travelers who attributed dramatic population shifts in Jordan and Tunisia to soil losses they wrongly believed had been caused by overgrazing.[5]

Montgomery's review of human societies depicts the typical cycle. Spanning the course of human history, he shows civilization after civilization settling in rich, fertile valleys, growing crops, then, as populations expanded, pushing crop farming up into the hills, often cutting down vast forests. But slopes, especially after being clear-cut and farmed in the same manner as valleys, quickly lose their virgin topsoils. Crop cultivation, we know, leaves part of the soil bare some or all of the time. This makes it especially "vulnerable to erosion, whether by wind or rain, resulting in a rate of soil loss tens to hundreds of times faster than nature makes it," notes Montgomery.[6] Whereas plants and the litter they produce protect the ground from the impact of raindrops and action of flowing water, on bare ground exposed to rain, the blast from each incoming raindrop sends dirt downslope.[7]

Over time, societies going down this path become incapable of feeding themselves. Much of the drive for Roman conquests, Montgomery argues, was fueled by poor agricultural practices that were whittling away the productivity of the empire's cultivated areas. Montgomery hypothesizes that exhaustion and erosion of the soil was a major factor in the fall of most once great civilizations, including the ancient Greeks, Romans, and Mayans. "[T]he human cost of soil exhaustion," he writes, "is readily apparent in the history of regions that long ago committed ecological suicide."[8]

There can be little doubt that since its beginnings some 12,000 years ago, agriculture has done inestimable damage to the globe. "Most analyses of problems *in* agriculture do not deal with the problem *of* agriculture," begins Wes Jackson's classic treatise *New Roots for Agriculture*. "[T]he plowshare may well have destroyed more options for future generations than the sword."[9] A bit later he continues: "So destructive has the agricultural revolution been that, geologically speaking, it surely stands as the most significant and explosive event to appear on the face of the earth, changing the earth even faster than did the origin of life."[10]

Speaking generally and of the United States, Wes Jackson wrote: "The hills are living, and, so long as they are clothed, eternal; the relatively flat lowlands, put to the plow by scarcely three generations of land stewards, are ephemeral."[11] American settlers followed the nearly universal pattern of failing to carry out the best management practices, especially the use of cover crops and the application of animal manure, that are necessary to keep soils generative and avoid erosion. While acknowledging that "[g]ood management can improve agricultural soils just as surely as bad management can destroy them,"[12] David Montgomery notes that "[d]ifferent types of conventional cropping systems result in soil erosion many times faster than under grass or forest."[13] The problem, Montgomery explains, is that soil organic matter declines under continuous cultivation as it oxidizes when exposed to air. Higher organic matter content means soils have more glomalin, and more soil aggregates, all of which inhibit erosion. "Thus, because high organic matter content can as much as double erosion resistance, soils generally become more erodible the longer they are plowed."[14]

These days it is often suggested that an important step toward correcting past missteps of agriculture would be reducing or removing animals—especially cattle and other grazing animals—from the world's food system. But as the previous discussions on carbon sequestration, soil, water, and biodiversity have illustrated, that would be moving in exactly the wrong direction. Humanity's greatest agricultural misdeeds have been carried out not with grazing animals but by ripping asunder the earth's dense protective plant cover.

The greatest asset of livestock, meanwhile, especially grazing animals, is that they can be raised for food without any plowing. They can be reared on grass, which, as we've already seen, benefits the world's climate and provides essential ecosystems for a menagerie of beasts from the tiny to the mega. Ecologically, even poorly managed rangeland is preferable to cropland. "Many ranchers do overgraze and their soil does erode, but even with overgrazing, poor ranching and no ethic, the land fares generally better when grazed than when put to the plow," Jackson notes.[15]

I readily acknowledge that livestock's impact on the earth's ecology has not always been positive. Here, the historical American West is a good case in point. Geography professor Paul Starrs, while appreciative of the important cultural and ecological role of American ranches, says in his book *Let the Cowboy Ride* that the forest reserve system was established in the early 1900s as a response to "virtually universal overgrazing in the Sierra Nevada of California and the mountains of northern New Mexico."[16] Starrs notes, "The years from 1880s to the mid-1930s marked a half-century of destructive exploitation of western rangeland that has few North American equivalents for intensity of rapine or unregulated voraciousness."[17]

However, this picture of cattle grazing is now out of date. As Starrs himself points out, more governmental regulations (starting with the 1934 Taylor Act, and continuing with environmental laws adopted in the 1970s) and a newly emerging scientific understanding of grazing's complex ecological impacts have led to much more carefully managed grazing in the United States.

In recent years, too, cattle numbers are substantially down. Although popular perception is that the United States has ever-rising numbers of cattle on the land, especially public lands, the reverse is true: Cattle grazing numbers have dropped significantly over the past half century. The federal Bureau of Land Management, which controls most of the grazing on public lands, reports that from 1954 to 2013 grazing use on public lands declined by 57 percent.[18]

A significant number of Americans have undoubtedly heard, or believe they have seen for themselves, how cattle grazing has caused environmental damage to America's West. Yet, like The Nature Conservancy's vernal pool studies, field studies measuring actual impacts have often surprised cattle's critics. One investigation of grazing's impacts by the USDA Forest Service found that removal of grazing in riparian areas might be beneficial for the short term, but would be "neutral to negative" over the long term. In part, the findings read:

[G]rasses have evolved with the periodic removal of vegetative material through fire, insects, or ungulates. In the

*absence of grazing or other disturbance, plants continue
to accumulate litter (dead grass blades left at the end of
the growing season). After years of litter accumulation,
plants go into a self-imposed stress whereby the detritus
(previous years' growth) chokes out new shoots competing
for light. The vigor of the entire plant is compromised and
rangelands become less productive and healthy. Many
invertebrate and wildlife species depend upon productive
grasslands, especially for winter range. In addition to loss
of plant vigor and decrease in rangeland health, the accumulation of litter allows fine fuels to build, which increases
susceptibility to fire.*[19]

Even more important, I believe it's a misnomer to label the
historical damage to America's West "overgrazing"—and one that
fosters fundamental misunderstanding. *Overgrazing* is a word fully
loaded with the ideas of too many animals, too much eating, too
much pooping, and too much trampling. I fully accept the point
that cattle can cause environmental damage to the land. But I
strongly reject the suggestion that anyone has proven the damage
to the American West comes from *too many* livestock. The land
has not been overgrazed, it has been *improperly* grazed. This is a
crucial distinction.

Why do I say this? We know that the earth was covered with
vast grazing, trampling herds of large animals for millions of
years. The 70 million bison extant in 1800 were just a remnant
of the populations of large herbivores once blanketing the North
American continent. We know, too, that pressure from abundant
predator populations kept the animals in close, dense, constantly
moving formations. We can see that the world's remaining large
wild herds—whether caribou in North America or wildebeest on
the Serengeti—still function this way today. The massive herd rolls
along the landscape as a single, giant organism—eating, pooping,
and trampling. Then the herd moves off and does not return for an
extended period of time. It was under such conditions that most of
today's plants, animals, and ecosystems evolved.

We also know that where similar conditions have been re-created in the modern world, dramatic restoration of land, water, and biodiversity has resulted. If we only allow our anti-cattle blinders to fall from our eyes for a moment, we can clearly see the logic, and the biological, historical, and real-world evidence. From this evidence we know it is not cattle, or grazing, or even overgrazing, but the way that cattle are managed today that is the problem.

Simply putting grazing animals out on the land cannot and will not have the same ecological effects. Because there are no longer robust populations of predators, which cause animals to bunch densely together and move as a herd, management techniques must be adopted to mimic the movement of historic herds. Environmentally beneficial stewardship of livestock is imperative. But overgrazing has very little to do with it.

Fortunately, some reconsideration is occurring—even in the mainstream—of the overgrazing concept. A 1998 report for the World Bank and United Nations was a harbinger of some of this shifting understanding:

> *Conventional wisdom suggests that much of the blame for "desertification" and land degradation in arid rangelands rests with pastoral livestock production. There is now considerable literature which corrects this misconception on two counts: the extent of dryland degradation is greatly exaggerated because underlying ecological dynamics have been misunderstood, and the contributory role of livestock has been mis-specified.*[20]

The shift in understanding has included increased attention to important factors like grazing intensity and timing. "While literature from the 1980s and early 1990s repeatedly linked livestock to the degradation of rangelands, more recent studies have refuted this by suggesting that prolonged rest leads to even more serious degradation," states a 2011 article in the journal *Pastoralism*.[21]

Lal and Stewart's major soil study notes, "Desertification is a biophysical process (soil, climate and vegetation) driven by socio-economic and political factors." Looking toward solutions, the paper offers that reversing desertification would "improve soil quality, increase the pool of carbon in soil and biomass, and induce formation of secondary carbonates leading to a reduction of carbon emissions to the atmosphere." To do this, it first recommends "establishing vegetative cover with appropriate species."[22] Presumably, this would include both climate-appropriate grasses and the animals needed to maintain them.

As greater evidence mounts that livestock, when properly managed, are not the problem, it is also mounting that they are essential to global environmental and food system solutions. In addition to the many benefits of grass areas on farms and grasslands, animal manure is increasingly being recognized as an asset the world food system cannot afford to be without. We've seen that adding manure to soil not only helps fertility but also reduces erosion. A University of Washington study that made direct measurements of soil losses between 1948 and 1985 found that a chemically fertilized field lost four times as much topsoil as an organically farmed field, fertilized only with manure.[23]

The hooves, mouths, and digestive tracts are even being increasing utilized to repair damaged land. In what's being called a cattle stomp, ranchers and land preservation groups in Colorado are working together for long-term regeneration of vegetation on a former mining operation's disturbed lands. Integrating biochar and compost and using rotational grazing, the stomp consists of native grass seed covered by a layer of wheat straw, where cattle are fed hay. The acreage is fenced off into small plots. As cattle graze hay in each plot, their hoof action "stomps" grass seed and straw into the soil, and their waste products provide moisture and natural fertilizer.[24]

From the historical to the current, real-world evidence shows that cattle grazing and trampling are not only beneficial, but in fact essential. In those situations where cattle and other grazing animals have caused environmental damage, the problem is one of improper management rather than overgrazing. For grazing to function as it should, it must be planned, and cattle must be actively managed. With such management, cattle can fulfill the essential ecosystem role of grazing and trampling herbivore.

CHAPTER 6

PEOPLE

THE PAGES OF THIS BOOK ARGUE that beef is healthy food, and that cattle, as herbivores that improve and add carbon to soils, and convert sunlight, in the form of grass, to meat and milk, are essential to a viable food system. Much has been said, and will continue to be said, on both sides about these hotly contested issues.

Less obvious and barely part of the public dialogue are some of the other benefits of keeping cattle. Namely this: We need *people* in agriculture, and cattle enable people to make a living doing it. Both varieties of bovines, dairy and beef, add value to diversified farms. Beef cattle also uniquely make possible, especially in the arid and semi-arid lands of America's Far West, the existence of homesteads and accompanying enterprises on remote non-tillable land, where crop farming is ill advised or impossible.

Being a professional farmer or rancher has never been an easy way to make a living. These days, it's particularly challenging. The U.S. Department of Agriculture reports that more than 93 percent of people involved in agriculture now have at least one other income source. Most people who actually work their own lands make, at best, a modest living. With those slim margins, animals are often essential.

I've met farmers raising diverse crops—cereals, vegetables, fruits, and livestock—who've told me that their farm's survival is owed to their animals, the most challenging yet most profitable aspect of the operation. Many of the ranches I know are in areas that are dry, windy, or hilly, or where rain falls only during the coldest part of the year. These are areas that have never been and never could be used to grow crops. Other stretches of the globe, including in the United States, have been so eroded and exhausted by plowing and crop cultivation that they can now support *only* grazing animals. All of these conditions work for grasses, and the animals that thrive on them, but not with cultivated crops. Cattle often render a living in agriculture financially viable for those who would be otherwise unable to make a living on the land.

Why does this matter? These days, only 17 percent of Americans live in rural areas at all and less than 1 percent are involved in agriculture. Some may argue that urbanization and the disappearance of family-scale farms and ranches are inevitable and all for the best. But I would strongly protest.

For one thing, we need more of the type of food that's generated at true family farms and ranches, not less. Food that's from known and trusted sources. Food that's more likely to be humanely produced, fresh, local, organic, sustainable, and not genetically modified.

More to my point, non-industrialized farms and ranches, especially those with animals, create an incomparable living environment for *humans*, one that is formative for both individuals and the nation. Because engaging in agriculture is one of the best ways for people to stay connected with the natural world, a country benefits by having a sizable portion of its population living and working on farms and ranches. As a nation, something intangible, yet vital, is slipping away as the number of farming and ranching families declines. It is diminishing our national character.

Daily, physical, outdoor labor that's connected to the land, dependence on weather and the elements, and the intense focus on the health and well-being of animals and plants all combine to make farming and ranching vastly different from living in the country while telecommuting to an urban job. Life on a farm or ranch strengthens bodies, enlivens minds, and enriches souls. Simply put, the farm environment cannot be created without farms. And it is a human environment eminently worth protecting.

On the dozens of farms I've visited and in the hundreds of farmers and ranchers I've met over the past decade, I've seen how a life working the land forms unique persons. They are characterized by resourcefulness, handiness, hardiness, physical courage, independence, and unflappability.

I've also discovered, sometimes a bit below the surface, a deeply embedded environmental ethic. These are people so immersed in the cycles and rhythms of life, they come to instinctively understand them. Mating, birthing, caring for young, growing, injury and illness, aging, and, perhaps most important, dying—all are part of the daily experience. They become second nature.

The hard work of farmers and ranchers, largely outdoors, is practiced by whole families, together. On the ranches I know the 3-year-olds have daily chores and 5-year-olds are on horseback; 13-year-olds drive trucks and 15-year-olds operate tractors. They are given important tasks with real responsibility; they contribute meaningfully to the family's collective work. And, invariably, the children rise to the occasion.

These children labor, yes, but they also play, and the play is grand. The great outdoors is their oyster, and nothing stands between them and countless adventures that are real, not contrived by overseeing parents or teachers. Through their own experiences, they learn both the patterns and vagaries of nature. Unlike today's urban and suburban kids, farm and ranch kids have hours, daily, to explore the natural world with minimal interference from adults.

At one ranch Bill and I visited, the eight-year-old son, after politely greeting us and chatting articulately for several minutes, excused himself with his fishing pole and sack lunch in his hand, his dog by his side. He was going, he explained, to his favorite fishing hole. He would be back before dinner. It was the first time I'd seen a child under 10 with that much independence since my youth.

At that time, the gap between farm and suburban life was not the chasm that it is today. Growing up in the outskirts of Kalamazoo, Michigan, in the 1970s and '80s, we were regularly on farms and had daily chances to play and explore out in nature. At least a couple of dozen times a year my mother and father took us to farms west of town, to pick fruits, buy eggs and vegetables directly from farmers, and, in December, to saw down our Christmas tree.

My daily dose of nature came closer to home. Just steps from our family's door was a 200-acre open parcel, a mixture of open grassy, shrubby, and wooded land. To many of the neighborhood's adults, this was a nondescript, undeveloped parcel. To us kids, it was a vast wilderness as well as home to countless insects, rodents, toads, deer, foxes, and birds. As many kids did in that era, I had broad freedom to roam. Part of nearly every day was passed there, climbing the trees, turning over rocks, following animal tracks, gathering flowers, seeds, leaves, and pinecones, and simply exploring. I remember lying alone on my back watching clouds float by. Sometimes I'd bring a book and would sit and read under a tree. Other times, I'd carry a backpack with an apple, a sketchbook, and some pencils. On many Sundays, our entire family would go there for a long stroll, then return home to warm up by the fireplace.

This seemingly undistinguished patch of land, bursting with life and constantly changing with the seasons, was my second home. I remember telling my mother that I was most certain that there was a God when I was out there in those woods and fields.

When I attended college in my hometown, majoring in biology, I studied cellular metabolism, ecosystems, and how plants and animals develop and grow. I enjoyed the fieldwork best. I returned to the plot of land I knew so well to gather my specimens. And for years to come, whenever visiting my parents, I went for a walk in those fields, usually with my father and his faithful canine companions. That plot of land, when later threatened with development, was the main reason I moved back from North Carolina to my hometown.

These days, growing up outside and falling in love with a particular wild place, the way I did, is a rare experience. In today's urbanized and suburbanized environments, children spend far less time outside, little time in nature, and virtually no time away from tight adult supervision. We think of this as the norm, and rarely consider the consequences of such an altered way of rearing our young. The sole exceptions to this societal norm today are kids who grow up on farms and ranches.

In his book *Last Child in the Woods: Saving Our Children from Nature-Deficit Disorder*,[1] journalist Richard Louv argues strongly and convincingly for the importance of youth spending substantial time in nature. Children themselves, and ultimately our country as a whole, he urges, are injured when children are divorced from nature, as so many are today. "If children do not attach to the land, they will not reap the psychological and spiritual benefits they can glean from nature, nor will they feel a long-term commitment to the environment, to the place,"[2] Louv writes.

Making occasional forays into parks and watching nature programs won't do the trick: "Passion does not arrive on videotape or on a CD; passion is personal. Passion is lifted from the earth itself by the muddy hands of the young; it travels along grass-stained sleeves to the heart."[3]

Living on our ranch for the past decade, where I moved from New York City, I've wallowed in the joy of returning to life amid the natural world, the same joy I had as a child. At the same time, I've gained perspective on the vast divide between urban and farm life, especially how it affects people's knowledge and connectedness with nature. We frequently host visitors here, usually chefs, environmental or animal activists, retailers, journalists, friends, or family. The degree to which they—adults and children alike—are unfamiliar with how nature works often becomes startlingly apparent.

I remember well a young woman visiting from Washington, DC, who was shocked and upset at seeing a nanny goat butting away a kid who tried to nurse from her udder. This was not a vicious butting. It was the rather typical *Get away from my udder* butt you might see any day in any group of goats or cattle. "Oh, that's normal. That's not her kid," I explained. When this did little to soothe our distressed visitor, I provided further explanation. "The mother needs to protect her milk for her own baby," I said, "and that kid has a mother of its own that he needs to go find." Nothing I said changed the opinion of our visitor that that was one mean goat.

I also recall a chef from New York City who blanched and looked ill when our Great Dane, Claire, proudly displayed a cherished prize in her mouth: a whole deer leg left over from a coyote feast.

I think back to an alert, happy, almost five-year-old boy visiting from San Francisco who was unfamiliar with the most basic biological facts. As we walked through our pasture surrounded by turkeys I began by speaking of the "males" and the "females." When I saw that wasn't resonating, I tried calling them the "boy turkey" and "girl turkey." Still nothing. Finally, I looked to the child's mother for help. "Maybe if you said 'mamma turkey' and 'papa turkey,'" she suggested. When that, too, failed to be understood I finally realized the boy had never grasped that animals, like people, come in two distinct genders.

The implications of our modern disconnectedness from nature and animals are manifold. As Richard Louv notes, it broadly affects attitudes. "To urbanized people, the source of food and the reality of nature are becoming more abstract. At the same time, urban folks are more likely to feel protective toward animals—or to fear them."[4]

Beyond that, I am increasingly persuaded that this disconnect is even making modern Americans more uncomfortable with the inevitable events of our own lives. Our lives will include aging and physical decline. All of us are sure to live through sickness, injury, pain, and, ultimately, death. Experiencing it through the lives of others gives us perspective that we are part of a much larger cycle of life, and it prepares each of us for our own journey.

Life on farms and ranches with livestock lends itself especially well to understanding life processes. Kids see, touch, hear, and smell plants and animals every day. Plants are sprouting, growing, blooming, fruiting, drying, dropping seeds, dying. Animals, both wild and domesticated, are mating, being born, nursing, growing, fighting, getting sick and hurt, and dying. Owls, ravens, hawks, coyotes, bobcats, gophers, rats; cattle, horses, dogs, cats, chickens, turkeys, goats, donkeys, mules. These are the ingredients of daily experience.

Part of every day is sure to involve observing wild animals (consciously or not). And much of the day will be spent caring for farm animals—providing feed and water, checking on them, searching for them, doctoring them, and, more likely than not, talking and singing to them. The needs, wants, and habits of creatures become intimately familiar to the farm child. Thinking beyond yourself becomes automatic.

Years ago, before I lived here, a young couple named Rob and Michelle Stokes lived here and did the day-to-day management of our land and our cattle. Now they have a place of their own in Oregon, and two young daughters. They raise cattle, goats, and heritage turkeys, all on grass. At one point, they were living back on our ranch and managing a large goat herd here, along with their daughter, three years old at the time.

One day Michelle asked for my help with a mother goat she'd been treating for pneumonia. At first, the goat seemed to have responded well to treatment. But that day she'd taken a turn for the worse and now was out in a pasture and too weak to stand. Michelle and I brought her back to the barn, laid her on some hay, out of the wind, and were trying to revive her with a liquid nutrient supplement. "C'mon, girl," Michelle kept softly urging her. She died a while later with her head in Michelle's lap. Michelle's daughter stood a few feet away, watching and asking questions. As Michelle worked steadily on the goat, she unwaveringly explained to her daughter what was happening, gently yet honestly answering all of her daughter's questions. I never forgot the skillful way Michelle handled her daughter's curiosity and concern.

Recently, Michelle, who now has two daughters, and I communicated about the value of raising children on a ranch, and it reminded me of the day we tried to save that nanny goat. She wrote:

My rancher children, in their short five and nine years, have seen more loss of life than most individuals in a life time and yet we still mourn these losses and understand loss. They are also blessed to be involved in the miracle of birth and the blessings of life over and over. They are witnesses to the struggles of saving and losing animals we love and they realize that there are many times that other lives take precedence over our own wants and sometimes our needs, truly giving them empathy for all living creatures.

Now that we have children of our own, I see, too, every day how much they gain from such an existence. A few days after his entrance into the world, our older son Miles began going out on the ranch with Bill or me, nearly every day. By two years, he played out in the yard for hours of most days. Out on the ranch, he comfortably ambled among turkeys that towered over him and eagerly begged to be useful.

We began giving him small, albeit real jobs like gathering turkey eggs with us, helping to herd the turkey flock (but not yet the cattle!), and filling waterers. At age five, he continues with those chores and also opens and closes gates, rounds up stray turkeys, and helps out in countless other ways. His help is valuable to us, and the pride he feels in his contributions is evident.

Now we have two boys. Like other ranch kids, they spend hours outside every day, breathing fresh air and soaking up the sun. They are surrounded by animals, of which they feel no fear. (We have to teach them when they *should* be afraid.) Our older boy can identify dozens of local birds, insects, flowers, and trees. Both of them love being physically active, especially outside.

In their remarkable book *The Geography of Childhood: Why Children Need Wild Places*, Gary Paul Nabhan and Stephen Trimble talk about the societal and cultural importance of children being reared so that they spend much more than occasional, carefully supervised time in natural spaces. Comparing their experiences to those of the Hopi in Arizona, they write about children reared on the ranch:

Ranch children learn in similar fashion about careful obser-
vation and nuance of behavior by taking care of horses and
pets and stock. They also understand domesticated animals
as utilitarian, existing to serve human needs. Death may
move them, but it does not shock them. Their intimacy with
animals is an intimacy with life.[5]

On our ranch, as on every farm and ranch that raises cattle, understanding nature is fundamental to the family's craft. In addition to paying attention to her animals, a good rancher knows the area's rainfall, wind, and temperature patterns. She grasps how they affect land, vegetation, and animals. She is knowledgeable about local wild flora and fauna; she is familiar with the area's rivers and streams and how they function. In short, a good rancher is not merely working outside. She is constantly interacting with— *depending on*—the natural elements in which she is functioning. It's the ideal incubator for true naturalists.

And we need naturalists now more than ever. Much more than just scientists who study natural ecosystems, true *naturalists* are people who love and understand the natural world in the tradition of Charles Darwin, Beatrix Potter, Henry David Thoreau, John Muir, Aldo Leopold, and Wendell Berry. Dr. Paul Dayton, professor of oceanography at Scripps Institution of Oceanography in La Jolla, California, points out that at the very moment when the earth is under its greatest assault academe virtually eliminated direct natural experience from its curriculum. This, Dayton urges, takes away "the opportunity for both young scientists and the general public to learn the fundamentals that help us predict population levels and the responses by complex systems to environmental variation."[6]

There can be little doubt that the environment will be preserved only if future generations understand and cherish nature. "If we are going to save environmentalism and the environment," Richard Louv writes, "we must also save an endangered indicator species: the child in nature."[7]

Farms, especially those with animals, have long been the perfect place for children to gain deep understanding of biology and to fall in love with nature and all its plant and animal components. Louv tells the following story of a longtime environmental activist:

> *When Janet looks back on her childhood in nature, she sees it not only as the source of her environmental activism—her work protecting the mountaintops of West Virginia—but also as nourishment for her own spirit. Her favorite place was a dairy farm run by her aunt and uncle . . . Off she would dash—to the barn, the henhouse, a hillside, meadow, or creek to explore the rich, natural treasure trove that lay before her. Whether she was watching the birth of newborn kittens or mourning the loss of a baby bird found featherless and cold on the ground, nature provided Janet with ample opportunity to feed her curiosity about life and taught her about the inevitability of death.*[8]

Besides being future environmentalists, people reared wholly or partially on farms and ranches become especially productive members of society. Before I started working as an environmental lawyer, I was only occasionally in the presence of cattle or the people who raised them. But I had been to quite a few small, diversified farms in southwestern Michigan. I'd seen enough to understand why Cato and Thomas Jefferson both held up the farmer as the ideal citizen. The farmers I remember from my youth were hard-working, doggedly independent, and civic-minded. They loved the land and had deep firsthand knowledge of the local climate and soils. They knew the area's flora and fauna, and were banks of local history lore. Of course they would be valued in society.

As Waterkeeper's senior attorney, I began working with farmers and ranchers from around the country, and spending more time on farms. Again and again, I was struck by the uniqueness of this set of people. And I began to see how the decline in the number of people in agriculture was more than just a national employment concern. It's costing us a pool of outstanding citizens with a special connection to nature and a unique set of character traits.

This is not just my own wild idea. Listening to a prominent CEO of a Fortune 500 company being interviewed on the radio one day, I heard something intriguing. The interviewer asked what the executive looked for in a job candidate. I expected his response to mention CVs padded with internships and Ivy League educations. But without hesitation he replied his first choice was always a person who grew up on a farm. Years of experience hiring and supervising people, he said, had taught him that farm kids were consistently dependable, hardworking, and resourceful.

Later, I saw a talk by a retired military officer who'd started a farm with his wife. Although he hadn't grown up on a farm, he said he wanted to raise his children on one. Decades in the military had taught him that the best servicepeople were those who came from farms. They were polite, disciplined, and reliable, which, he said, was just how he wanted his kids to turn out.

I suspect many college admissions directors would view things similarly. My parents, who both spent their careers teaching at universities, said farm life seemed to form especially good students.

Statistics showing the disappearance of farmers and ranchers are often recited and lamented. Yet this aspect of America's shifting demographic has been largely ignored. As we lose farms and ranches, we are losing an irreplaceable environment for child rearing and an incubator for good citizens.

Equally underappreciated is that we are losing a physical environment that creates strong, healthy children.

As our son Miles was still in the womb, I did several hours of chores on the ranch, every day. Clearly, I had no choice but to bring Miles with me. (Well known among friends and family is the infamous story of my water breaking out on the ranch, during morning chores, followed by me attending Easter Sunday services and singing a Bach solo before heading to the hospital.)

Sometimes I wondered whether it was such a good idea to expose my unborn child to the dust from turkey feed, hay, manure, and dander from cattle and turkeys. I should not have worried. New research is turning my concern on its head. Rather than harming my sons, my working around animals has likely been protective of their health, as will be their experiences growing up with livestock.

A 2013 *New York Times* op-ed explains why a growing number of scientists consider children who grow up on farms with livestock to be the best protected from allergies and asthma.[9] Part of the research was triggered by the discovery that the Amish, 92 percent of whom often spend time on farms, have extremely low rates of allergy and asthma. "Westernization," or maybe more accurately "industrialization" (which the Amish have strongly resisted), has been connected with people becoming allergic. Until fairly recently, it was not understood why.

Now it is believed that the so-called farm effect protects farm kids by stimulating their immune systems. Humans on farms and ranches, beginning even before birth, are chronically exposed to high levels of microbes, such as from cattle manure, dander, and hay. The earlier the exposure starts, including during pregnancy, the greater the protection. "Children born to mothers who work with livestock while pregnant, and who lug their newborns along during chores, seem the most invulnerable to allergic disease later," the article notes. Research by Bianca Schaub, a doctor and researcher at Munich University, shows that newborns from farms are better at quashing allergic-type reactions. "The more cows, pigs and chickens a mother encounters, essentially, the more easily her offspring may tolerate dust mites and tree pollens."

The research also suggests that mothers on livestock farms and ranches might provide a natural therapy to the baby's immune system, "one that pre-programs the developing fetus against allergic disease."[10] Increasingly, science supports the conclusion that recent decades' epidemic levels of allergy and asthma are linked to our modern-day disconnect with farms, particularly farm *animals*.

The idea is not actually new. Jared Diamond's brilliant book *Guns, Germs, and Steel: The Fates of Human Societies,* first published in 2006, includes the argument that historical events demonstrate the importance of exposure to livestock, especially cattle. Dissecting the complexity of reasons certain cultures became more technologically advanced and, consequently, more politically and militarily dominant over others, he argues that domesticated animals, especially cattle, helped humans develop immunities to infectious diseases.

By geographic accident some cultures arose in regions where animals existed that were prone to domestication. Cattle were particularly significant because they could pull carts and plows, as well as provide meat, milk, and manure. Living in close proximity to domesticated animals, humans exposed themselves to more disease-carrying organisms. On the one hand, this contributed to epidemics. On the other hand, Diamond argues, coming in contact with the germs carried by cattle and other animals was critical to the successes of those cultures. By being exposed to the diseases, these populations developed immunities.

However, when those societies collided with cultures that had no domesticated animals, and therefore had not been exposed to livestock-borne diseases, the results were catastrophic for the cultures without livestock. The vast majority of the deaths wrought by Europeans in the New World were from their germs—germs to which the native peoples of the Americas had never previously been exposed. The spread of germs, and diseases, massively abetted the military defeats inflicted upon Native American cultures throughout the American continents.

In addition to the surprising benefits of being around lots of germs, the farm kids I know (including our own) tend to spend much more time outdoors than glued to a computer or TV. The typical American child now spends an average of three hours a day in front of a television and five to seven hours a day in front of a screen (which includes time playing video games, looking at Facebook or Twitter, and otherwise using a computer).[11]

Although most kids start the habit early, the American Academy of Pediatrics recommends that children two and under have *zero* "screen time." Allowing young kids to watch TV and videos has been linked to various behavior problems, especially attention deficit hyperactivity disorder (ADHD).[12] The increased amount of time children spend staring at screens has also been linked to the rise in obesity and vitamin D deficiency.[13]

"Several large-scale studies have found that vitamin D deficiency is widespread—one in ten U.S. children are estimated to be deficient—and that 60 percent of children may have suboptimal levels of vitamin D," states the website of the Johns Hopkins Children's Hospital.[14] Normally, 90 percent of our vitamin D comes from time spent in sunlight.[15] Getting sufficient vitamin D from foods is difficult, and research suggests that vitamin D pills are not equivalent to sunshine. Being outside every day is vitally important.

These issues—the physical environment and quality of ranch life, especially for children—may seem remotely related to cattle. But they are real and important. Cattle make a lot of American farms and ranches possible. If cattle disappeared from our food system, these human environments would disappear as well. And that's a loss to our culture and society we cannot afford and should not allow.

BEEF

FOOD AND HEALTH

CHAPTER 7

HEALTH CLAIMS AGAINST BEEF:

The Rest of the Story

FEW DOUBT THE EMERGENCE IN RECENT DECADES of a public health crisis connected with the American diet. Clear and shocking statistics show a country whose people are getting fatter, have higher blood pressure, and are increasingly suffering and dying from chronic diseases, including heart disease, diabetes, and stroke.

At the center of the storm is our widening girth. Two-thirds of American adults (69 percent) are now overweight, with 39 percent obese (men having more than 25 percent body fat, and women with more than 30 percent). From 1960 to today, the average weight of American men rose from 166 pounds to 191 pounds; that of women went from 140 pounds to 164. So many human health problems are triggered by carrying too much body weight that the dramatic rise in obesity is cause for great concern.

For young people, the situation is graver still. When I think back to my childhood, I can remember just one child, in a neighborhood full of kids, who would meet the definition of obese. However, from 1970 to 2000, the number of overweight children more than doubled, and the obesity rate tripled. Nearly one-third (32 percent) are now overweight, with 17 percent obese. The American Public Health Association recently averred that the situation is "leading to a generation at risk for cardiovascular diseases, diabetes and other serious health problems."[1] Dr. Robert Lustig, a physician at University of California–San Francisco who has clinically treated and studied obese children for the past two decades, reports that it is even affecting babies. "We now have an epidemic of obese six-month-olds," Lustig warns.[2] (Some of the possible reasons why babies are affected will be discussed a bit later.)

Closely related to the trend of heavier weights is the rise in chronic conditions and diseases. One-third of U.S. adults (33 percent) now have hypertension. The four leading causes of U.S. deaths are *all* chronic diseases.[3] In order from first to fourth, those are: heart disease, cancer, respiratory disease, and stroke.

The trends are particularly unsettling. Obesity, hypertension, and chronic diseases are increasing or merely leveling off despite several factors that *should* have dramatically reduced them. Take heart disease and its related deaths. Advanced medical interventions for heart attacks are available at every hospital; stent insertions and coronary bypass surgeries have become routine. Authors of a 2011 *Journal of the American Medical Association* article on heart procedures noted: "Coronary revascularization, comprising coronary artery bypass graft (CABG) surgery and PCI [stent insertion], is among the most common major medical procedures provided by the U.S. health care system, with more than 1 million procedures performed annually."[4] At the same time, smoking, among the top risk factors for heart disease, is much less common. Adult male smoking rates plunged from 55 percent in the 1950s to 18 percent today.[5]

Nonetheless, heart disease incidents and death have merely leveled off in the past decade rather than declining. Today, it is responsible for 42 percent of adult deaths, almost four times the rate of 1900, when heart disease is believed to have caused just 8 percent of deaths[6] and was the fourth leading cause of death.[7] In real numbers, far more Americans are dying from heart disease than ever.

Physical activity is undeniably another important factor in modern chronic diseases. With the advent of cars, televisions, and computers, Americans became far more sedentary. While today 20 percent of people have jobs that involve strenuous physical labor, in the mid-20th century it was around 50 percent.[8] Globally, physical inactivity has had as deleterious an effect on health as smoking and obesity, says Dr. I-Min Lee, a professor of epidemiology at the Harvard School of Public Health, who led a recent study on the effects of inactivity.[9] However, research also suggests that much of the U.S. shift toward sedentary lifestyles had already occurred by 1960 or 1970.[10] So lack of exercise certainly does not fully explain the latest, worrying trends.

Why in recent decades has an obesity epidemic emerged, and why do heart disease and other chronic diseases continue to plague us despite changes that should have dramatically reduced them? The evidence points to diet. But not in the ways you might expect.

How Our Diet Has Really Changed

For the past 50 years we've heard repeatedly that diet is the linchpin to modern Western health problems. Author Michael Pollan has aptly described the situation this way: "Scientists operating with the best of intentions, using the best tools at their disposal, have taught us to look at food in a way that has diminished our pleasure in eating it while doing little or nothing to improve our health."[11]

Over and again we've been told a single, simple message: What's wrong with our diet boils down to the butter on our toast, the cream in our coffee, the steak we're having for dinner. Nearly universally, public health agencies, dietitians, and doctors have advised us to cut back on red meat and fats, especially animal fats.[12]

The origins of this advice can be traced back to a 1953 study often called the Seven Countries Study,[13] conducted by Ancel Keys, a University of Minnesota epidemiologist. Keys believed that saturated fat was the primary cause of heart disease. The idea made him famous (Keys graced *Time* magazine's cover in 1962) and changed the eating habits of generations to follow. Even without ever having heard of Keys in my youth, I'd heard his idea so often that I considered it incontrovertible truth in college, when I decided to adopt a vegetarian diet.

Millions of others also took the message to heart. For the second half of the 20th century, especially after 1970, Americans followed the advice they were hearing everywhere and cut back on red meat and animal fats. We cooked and baked with vegetable oils. (I clearly remember my parents switching from sticks of butter to tubs of corn oil margarine.) Americans also dutifully ate more fruits, more vegetables, and more whole grains. Most people I talk to today still consider it gospel that healthy eating means avoiding fats and red meats.

Yet over those decades, America's dietary shift did not improve its collective health. In fact, it got considerably worse. Obesity and hypertension soared, and heart disease and stroke rates persisted. The advice we heard for decades—that reducing fats and red meats in our diets would improve our health—was proving wrong.

Partly because this colossal failure is now so readily apparent, medical and public health researchers and practitioners have been reexamining the standard dietary advice. And a sea change in thinking is under way. A few brave doctors and scientists have begun publicly acknowledging that Americans have been getting bad dietary advice for the past half century.

Momentarily, I'll discuss in some detail what the medical and public health research really shows about linkages between various things we eat and drink and the current health crisis. But first, it's important to dive into some specific data about how our eating patterns have *actually* changed over the past century, especially in recent decades.

Consumption levels of just about every food you can imagine can be determined by looking at relatively undisputed government data. For over a century, the U.S. Department of Agriculture (USDA) has tracked the amounts of everything we eat and drink.

The first point to note is that Americans are simply eating a lot more these days. Between 1970 and 2003, USDA research estimates that the average daily caloric intake increased by a staggering 523 calories per person.[14] Theories abound as to the cause of this increase. Regardless of the reason, this fact alone certainly explains part of the problem.

Also troubling is the precise source of those additional 523 calories. Fast food now accounts for a third (34 percent) of the calories Americans consume. And then there's snacking, which accounts for more than half of the recent calorie increase.[15] The amount we take in at dinner has actually declined. In other words, more Doritos, fewer rutabagas. Or perhaps, more accurately, more doughnuts, less steak.

Which leads to an important first question: Are we really eating more red meat and animal fat than ever?

In a word: no. In fact, the reverse is true. We are eating less of both.

Throughout the course of the 20th century, meat consumption fluctuated. Due to wartime rationing, there was a dip in the 1940s. (Sugar was also rationed, which, as will become clear shortly, is noteworthy.) Around 1970 (most likely due to uncharacteristically low prices at the meat counter resulting from overproduction), there was an upward bump of beef eating. But over the course of the century, and especially for the past three decades, the general trend for both red meat and animal fats was downward. Both are notably lower today than they were a century ago.

Specifics are helpful here. (All of the figures for farming and food production and consumption, except where otherwise indicated, are from electronic versions of original U.S. Department of Agriculture records, which I have personally examined.) While Americans ate 71 pounds of beef per person per year in 1905, they ate 60 in 2010. In those same years, veal consumption went from 7 pounds per person to just 0.4 pound. Lamb consumption went down from 6 pounds to just 1. And pork consumption went from 62 pounds to 48. In a similar pattern, egg consumption went from 284 per person to 243. All significantly down. Stated another way, compared with a century ago, we are now eating 11 pounds less beef per year, 6.5 pounds less veal, 5 pounds less lamb, 14 pounds less pork, and 41 fewer eggs.[16] These are hardly the statistics of a nation that has increased its animal fat and red meat consumption.

Now let's look more closely at recent decades. In that time period, red meat consumption went strongly down, especially beef. From 1970 to 2005, beef consumption decreased by 22 percent; pork went down by 3 percent. Overall, red meat consumption decreased by 17 percent.

For foods rich in saturated fats, the trend is even clearer. From 1970 to 2005, butter consumption declined by 15 percent; lard went down by 47 percent; and whole milk consumption plummeted by 73 percent. The only major exception to this pattern was cheese (from pizza and other fast-food consumption), which increased.

The rise in cheese eating, however, was not enough to counteract an overall consistent shift toward less consumption of other animal fats. Mary Enig, a PhD food scientist specializing in fats who authored *Know Your Fats*, has calculated that over the course of the 20th century Americans' overall consumption of saturated fats went down by 21 percent.[17]

At this point, we may well wonder how, in the face of these consumption facts, anyone could blame America's health crisis on animal fats and red meat. Indeed, if we believe in the Keys hypothesis, this data is deeply confounding.

But things get still sketchier. Changes in red meat and fats are only part of the story of our shifting eating and drinking patterns. As our consumption of animal fats went down, our vegetable fats went way up. Total per-person cooking and baking fats more than doubled from 1909 to 2009, from 36 pounds to 80 pounds, increasing by 63 percent from just 1970 to 2000. From 1910 to 1970, vegetable fat consumption had already gone up by over 400 percent.[18] A lot of this change was certainly the result of Americans attempting to follow health advice that they should replace animal fats with vegetable fats. Vegetable fats tend to cost less, too, which has certainly been another factor in the shift.

Similarly, as physicians and nutritionists advised us to get away from red meat Americans dutifully migrated away from pork, beef, veal, and lamb, and began eating substantially more chicken, turkey, and fish. In 1909, Americans ate 15 pounds of chicken per year; by 2009, that had gone up to 80 pounds, a fivefold increase. Average turkey consumption went from 2 pounds in 1935 to 9 pounds in 2009, a fourfold increase.[19] Fish consumption also rose considerably. From 1910 to 1999, per-capita consumption rose from 11.2 to 15.3 pounds per person per year (a 37 percent increase).[20]

If red meats are the problem, and poultry and fish are so much better for us, why have all the major diet-related diseases gotten worse as we've eaten more of the "white meats" and cut back our lard, dairy fat, and red meat?

And what about those vegetable fats? Aren't they supposed to be so much healthier than fats from animals? That advice may turn out to have been the worst of all. For one thing, a lot of the vegetable fats Americans used in the 20th century were partially hydrogenated vegetable fats—in other words, the odious man-made trans fat.

"Trans fat is formed during food processing when some hydrogen is added to vegetable oil to increase solidity," explains an article in *Scientific American*. "It is typically added to food to increase its shelf life, improve its texture and maintain its flavor."[21] Woo hoo! Long shelf life! Then it can sit in a warehouse or vending machine for months before we eat it! Great.

Crisco kicked things off in 1911, introducing a shelf-stable fat for baking. It was the first time artificial trans fat came into the American diet. Over the 20th century, trans fat became a ubiquitous staple, found in everything from potato chips to Oreos and Girl Scout cookies. I vividly remember TV ads in my childhood featuring Loretta Lynn hawking Crisco for "the flakiest pie crusts."

(As an added bonus, it needed no refrigeration and could sit in your cupboard for months.)

Ironically, consumer advocacy groups successfully lobbied fast-food restaurants to replace beef tallow and tropical oils with partially hydrogenated oils containing trans fat. McDonald's famously stopped using beef tallow for its french fries. It was immediately obvious to many that this lessened the foods' palatability. In hindsight, it was also injurious to public health.

Due to scientific research connecting it to heart disease, Denmark began strictly limiting manmade trans fat in foods in 2003. Other European countries and some Canadian cities and provinces followed suit. When Mayor Bloomberg pushed through a trans fat ban for New York City restaurants in 2006, some of my New York friends said he was a crusading public health zealot who was stepping over the line.

But history is already vindicating Bloomberg. Earlier that year, the U.S. federal government had begun requiring labeling of foods containing artificial trans fat. And in November 2013, the FDA tolled the death knell for manmade trans fat by announcing that it should eventually be removed from foods entirely. In so doing, the FDA noted the U.S. Centers for Disease Control and Prevention had estimated that getting rid of trans fat would prevent an additional 7,000 deaths from heart disease annually and up to 20,000 heart attacks every year.[22] "There can be little doubt that the presence of laboratory manufactured trans fats is part of reason for the rise in chronic diseases," said the FDA announcement. Noting that part of its responsibility is to ensure a safe food supply, the FDA announced it had made a preliminary determination that trans fat no longer met the standard of "generally recognized as safe." If its preliminary determination is finalized, foods containing artificial trans fat would be considered adulterated and will become illegal.

Unfortunately, the government's action, while laudable and correct, is a century too late. Tens of thousands of people who ate cookies, pies, and french fries containing artificial trans fat made from vegetable fats now suffer, or one day will suffer, cardiovascular disease and heart attacks as a result. Clearly, trans fat partly explains why heart disease rates rose over the 20th century and have stubbornly refused to go down.

It's worth noting that the United States' leading researcher on the health effects of trans fat, Dr. Fred Kummerow, who warned of their dangers decades ago, is one of numerous prominent scientists who have long and vociferously questioned Ancel Keys's hypothesis that saturated fats cause heart disease. Despite the recent acceptance of Kummerow's warnings on trans fat, he still considers the public health message about fats totally wrong.[23] Kummerow has been steadily conducting research on fats since he was a young nutrition PhD scientist at University of Illinois in 1957. Still actively researching at age 99, he says there is nothing inherently unhealthy about the saturated fats found in foods from animals. (And he eats eggs fried in butter every morning for breakfast.)

What matters, he says, is whether the fat is oxidized. "Cholesterol has nothing to do with heart disease, except if it's oxidized," Dr. Kummerow said in a 2013 *New York Times* interview. High temperatures used in commercial frying cause polyunsaturated vegetable oils to oxidize, while soybean and corn oils can oxidize inside the body. "If true," the *Times* article notes, "the hypothesis might explain why studies have found that half of all heart disease patients have normal or low levels of LDL [low-density lipids, often referred to as "bad cholesterol"]."[24] (Interestingly, as I'll discuss later, very-high-temperature food preparation may play a substantial role in turning otherwise healthy foods into foods that, over time, cause health problems.)

The most troubling aspect of the whole trans fat tragedy is that it was so unnecessary. Tens of thousands of people have almost certainly suffered and died because they ate artificial trans fat. Yet none of it needed to happen. Trans fat served no noble purpose like making food more nutritious or healthier (or even tastier!); nor was it created to feed starving people. Rather, it was simply a laboratory-produced substance to give food added shelf life. Is it any surprise that such an item would make us sick?

The situation fits perfectly into a general rule suggested decades ago by British physiologist, professor of medicine, and clinical physician Dr. John Yudkin (about whom I will say more later). Dr. Yudkin spent his career conducting epidemiological and clinical research on the effects of various foods on the human body. His particular interest was heart disease. Decades of investigation led him to reach the following eminently sensible conclusion: Any food that humans and their ancestors have been eating for millions of years is unlikely to cause heart disease or other serious, chronic diseases. He considered meat, fish, fruit, and vegetables to be among the inherently *un*suspicious foods.

The converse, Yudkin urged, is equally true: Any food that humans and their ancestors did *not* eat during their millions of years of evolution should be considered suspect as to its impacts on human health.[25] Sucrose (sugar) is an example of such a food. Such a simple, sensible idea. Applying Dr. Yudkin's principle, laboratory-generated trans fat should not have been widely introduced into the food chain until long-term studies had proven it safe. Had we taken such an approach, the lives and health of tens of thousands of people would have been spared.

Yudkin's idea, that foods our bodies evolved with are innocent until proven guilty, may have been deemed quaint by some of his peers in the 1960s and '70s, as space-age foods like Tang and sugar-loaded treats like Twinkies were being invented. Today, however, it is being taken seriously by leading scientists and physicians. Momentarily, I'll discuss Dr. Yudkin's main findings and recent research that backs them up.

Doubts and Proofs

First, though, I want to return to the overall question of proof that actually exists to connect red meat and saturated fats to the current epidemic of obesity, hypertension, and chronic diseases. There are several broad categories of studies used to determine what causes chronic diseases.

The consumption data we've been reviewing would be used in epidemiological studies. So far, we've seen quite clearly that, compared with a century ago, Americans are eating less saturated fat and less red meat and, instead, eating more vegetable fat, fish, and poultry. USDA consumption data also shows we've been eating more whole grains, fruits, and vegetables over the past three decades. Additionally, we've seen that various other factors that significantly affect chronic disease rates and their complications (smoking rates, medical care) have changed in ways that should lead to lower rates of disease and fewer deaths. Yet all of these positive changes have failed to counteract the health trends, at least in a meaningful way, which is quite alarming.

On their own, of course, food consumption data could never be used to prove that one food or another is causing America's health problems. Such data helps devise theories; it helps draw up a list of suspects; it's one piece of the puzzle. As a former assistant district attorney, I know all too well that in a courtroom you must prove a defendant's guilt beyond a reasonable doubt. A defendant, meanwhile, never needs to *prove* his innocence. The accused merely needs to pose unanswered questions—to raise doubts—about his guilt.

As a trained biologist, I am also familiar with the methods used to test a scientific hypothesis. As in law, declaring a hypothesis sound carries a much heavier burden than merely poking holes in it. There is (or should be, anyway) a high bar for concluding that a food causes chronic diseases. Taking a cue from Yudkin's idea, I would add that the burden of proof should be especially high when the food in question has been part of the human diet for millions of years. While it can be nearly impossible to prove with absolute certainty the cause of any long-term illness, there are various forms of evidence that, in combination, can reasonably be deemed proof.

If red meats and animal fats now stand accused (and there's little doubt that they *do*), and Ancel Keys is their original accuser, it's easy to see here that the consumption data raises serious doubt about his hypothesis. Said another way, the consumption data would be insufficient to affirmatively *prove* that red meat and fats are good for you, but they are ample to raise serious *doubt* about whether those foods are guilty of causing America's poor health, as has been so commonly alleged.

Taking this analogy a bit farther, the Keys "Seven Countries" study could never meet the standard of proof needed to establish the guilt of red meat or saturated fat. Nor, from a strictly scientific view, is it anywhere near sufficient to prove his hypothesis. Keys merely looked at two conditions that existed in the same country (collective saturated fat consumption levels and heart disease rates), noted a correlation, then inferred causation. This falls squarely into the category of epidemiological study, research that should trigger various forms of additional investigation into a hypothesis.

To be sure, that does not suggest that observational, ecologic research like Keys's study is worthless. It can be invaluable in certain situations—if, for example, you were trying to figure out from where an infectious disease outbreak is originating. Even without interviewing the afflicted, merely plotting their physical locations could be crucial in coming up with a list of suspect sources for the pathogen. In contrast, when attempting to prove causation for a chronic disease from a particular food item, Keys's type of epidemiological research is weak and should only ever have been a starting point.

Here's why. Keys showed nothing whatsoever about the diets of the individual people in the seven countries. He had no way of knowing whether *any* of the people, let alone *all* of the people, eating higher levels of saturated fats were the same people who were getting sick with cardiovascular disease. Theoretically, it is entirely possible that *none* of the people eating large amounts of saturated fat had heart disease, just as it is possible that people who ate very little saturated fats were those with the disease. All Keys showed was that in seven countries the *total amount* of saturated fat consumed correlated, to an extent, with the *total presence* of heart disease.

Keys's conclusion is further weakened by the fact that the same type of data he used for the 7 countries was available for at least 22 countries. He used only data from the seven countries that most strongly supported his thesis.[26] When data from all 22 are graphed, the correlation appears weak indeed. For all these reasons, the Keys study, on its own, is poor-quality evidence. Certainly, it is insufficient to convict saturated fat of the crime of causing America's current epidemic of poor health.

The other major piece of evidence taken as proof of the saturated fat hypothesis relates to cholesterol. Clinical trials in the 1970s, following publication of Keys's theory, showed that eating saturated fat raises levels of low-density lipid cholesterol (commonly referred to as LDL cholesterol, or just LDL). This phenomenon is now well documented. As I'll describe in more detail momentarily, it is now recognized that such research has serious limitations. In the early days of cholesterol research, however, few nuances were understood. Saturated fat's effect on LDL was deemed sufficient proof that it was guilty of causing coronary disease, and the matter was treated as a closed case.

Lately, however, the case against saturated fat and red meat is being reopened. Three things seem to have changed. First, it has become increasingly apparent that there were problems with the original research, such as those just described. Second, there have been serious flaws underlying subsequent studies that seemed to link beef and saturated fat to health problems. And third, as I'll talk about more in a bit, new research (and some older research) links chronic problems to other foods.

One serious shortcoming is what's called the healthy user bias. Chris Kresser, an author and integrative medicine practitioner in Berkeley, California, who considers both beef and fat important components of a health-promoting diet, explains it like this: "Since red meat has been vilified for years in the mainstream press, people who eat less of it are also more likely to eat less of other foods that *are* actually unhealthy (i.e. refined sugar, transfats, processed foods, etc.) and to engage in healthier lifestyle choices (i.e. they are physically active, don't smoke, etc.)." Kresser also points out that the types of questionnaires used to track food intake are notoriously unreliable. "Do you remember what you ate for lunch last Tuesday? Neither do I." Combined, such flaws in the research have created a body of unreliable epidemiological studies on red meat.[27]

Additionally, numerous major, credible studies have found no link between saturated fats and/or red meat and human diseases. At the same time, a growing body of research is documenting that things other than fat and red meat are more likely to be responsible for today's health problems.

To start, here are some of the studies and re-analyses of the data that have chipped away at the widespread acceptance of the claim that red meat and saturated fat cause heart disease.

A 1998 meta-analysis published in the *Journal of Clinical Epidemiology* reviewed dozens of studies including ecological, dynamic population, cross-sectional, cohort, and case-control studies, as well as controlled, randomized trials of the effect of fats in the diet.[28] In particular, it examined the role of saturated and polyunsaturated fatty acids in cardiovascular disease, noting that a diet rich in saturated fats and low in polyunsaturated fats "is said to be an important cause of atherosclerosis and cardiovascular diseases (CVD)." For each study type, the review found highly contradictory study results. Even the type of evidence made famous by Ancel Keys—positive correlations between national intakes of total fat and saturated fat and cardiovascular mortality—were "absent or negative in the larger, more recent studies," the analysis concluded. Overall, the study found a lack of evidence linking saturated fat and coronary disease in any type of study.

This was followed by a 2010 study published in the *American Journal of Clinical Nutrition* and conducted by physicians at the Oakland Research Institute. It was a meta-analysis of 21 studies (including 347,747 subjects) to evaluate the association of saturated fat with cardiovascular disease. Noting that "a reduction in dietary saturated fat has generally been thought to improve cardiovascular health," the research aimed to "summarize the evidence related to the association of dietary saturated fat with risk of coronary heart disease (CHD), stroke, and cardiovascular disease (CVD; CHD inclusive of stroke) in prospective epidemiologic studies." Using the 21 studies and a random-effects model, it derived composite relative risk estimates for CHD, stroke, and CVD. Finding no credible evidence of a link, the study authors concluded: "Intake of saturated fat was not associated with an increased risk of CHD, stroke, or CVD."[29]

Also in 2010 came a major study by researchers from the Harvard School of Public Health,[30] who found that although eating processed meat (like bacon and baloney) was associated with a higher risk of heart disease and diabetes, there was *no increased risk at all* from eating *unprocessed* red meats, including beef, pork, and lamb. Researchers noted that while dietary guidelines tend to recommend reducing meat consumption, "prior individual studies have shown mixed results for relationships between meat consumption and cardiovascular diseases and diabetes." The reason for the mixed results, the Harvard team concluded, was that prior studies had failed to separately consider the health effects of processed versus unprocessed red meats. The team began by systematically reviewing nearly 1,600 studies. Twenty relevant studies were identified, which included a total of 1,218,380 individuals from 10 countries on four continents. Neither the fat nor the cholesterol content of the processed meats varied from the unprocessed, the researchers pointed out. Rather, the key difference was the sodium and nitrate content, leading the study to conclude: "[D]ifferences in salt and preservatives, rather than fats, might explain the higher risk of heart disease and diabetes seen with processed meats, but not with unprocessed red meats."

If any doubts on the question lingered, the final nail in the coffin of the alleged link between saturated fat and heart disease came in early 2014. In what *The New York Times* deemed "a large and exhaustive new analysis by a team of international scientists," a report published in the journal *Annals of Internal Medicine* found "no evidence that eating saturated fat increased heart attacks and other cardiac events."[31]

The researchers based their conclusions on nearly 80 studies involving more than half a million people. They looked at what people reportedly ate, as well as more objective measures such as the composition of fatty acids in their bloodstreams and in their fat tissue. Additionally, the scientists reviewed evidence from 27 randomized controlled trials (often referred to as "the gold standard" for such research) that assessed whether taking polyunsaturated fat supplements, such as fish oil, promoted heart health.

Combining these investigations, the researchers concluded that saturated fat consumption had no connection with heart attacks and that higher consumption of monounsaturated fats (like olive oil) and polyunsaturated fats (like corn oil) did not lessen heart disease risk. "The new findings are part of a growing body of research that has challenged the accepted wisdom that saturated fat is inherently bad for you," the *Times* article continued. Summarizing the findings of this research, Dr. Rajiv Chowdhury, a cardiovascular epidemiologist at Cambridge University and the lead author of the study, said: "It's not saturated fat that we should worry about."

These and other similar studies exonerate beef and saturated fat from accusations that they cause heart disease. When considered together with the flaws in the studies that purported to show negative health effects of beef and saturated fat, the evidence is extremely compelling that neither red meat nor saturated fat is bad for your health. The evidence also suggests that additives frequently used *with* meat are to blame for health problems often blamed on meat itself. Meanwhile, the way meat is cooked may also contribute to associations found by some studies between meat and various health concerns.

Famed food scientist Harold McGee advises: "We should . . . prepare meat with care." McGee points out that, in addition to nitrosamines that can be created by very high heating of nitrate-containing meats, scientists have identified two other families of potentially carcinogenic chemicals that can be created by popular cooking methods. One is heterocyclic amines (HCAs), which can be formed at high temperatures by the reaction of creatine and creatinine (compounds present in small amounts in meat) with amino acids. Second, polycyclic aromatic hydrocarbons (PAHs) are created when organic material is heated to the point that it begins to burn. Cooking over a smoky wood fire therefore deposits PAHs from the wood onto the meat (as it would, it should be noted, on any food cooked over an open flame).[32] Beef that is unprocessed, not burned, and not prepared over an open flame raises none of these concerns.

But what about the effect saturated fat has on LDL cholesterol levels? As previously mentioned, clinical trials have shown that saturated fat raises LDL. In the 1970s, this was considered highly damning. Now we know better. The current understanding of cholesterol is far more nuanced and multidimensional.

For one thing, as already mentioned, some 75 percent of heart attack patients have normal or low LDL cholesterol levels.[33] This fact alone establishes that there must be much more to heart disease than elevated LDL levels. Moreover, it is now widely recognized that a connection between saturated fats and LDL levels is not proof of a link to *coronary disease*. Plainly stated, having an elevated LDL is not an illness. Rather, LDL levels are simply biomarkers—things doctors use to help predict whether or not someone might develop a disease. And the LDL number is just one part of the information a medical professional needs for a patient's cholesterol numbers to serve as a useful biomarker. Levels of HDL (high-density lipids), the other major type of cholesterol, and the ratio of LDL to HDL are now considered at least as important.

Clinical investigations into the effects of diet on LDL and HDL levels have revealed a complicated picture. Researchers in the Netherlands worked to clarify it by reviewing results from 60 human dietary clinical trials.[34] Noting that "the ratio of total to HDL cholesterol is a more specific marker of CAD [coronary artery disease] than is LDL cholesterol," the study evaluated the effects of individual types of fats on cholesterol ratios. It did find that saturated fats increased LDLs, but, as Dr. Dariush Mozaffarian, cardiac epidemiologist and faculty member at the Harvard School of Public Health, said of this research, "not by much." Equally important, the study also concluded that if *carbohydrates* replace dietary saturated fats, the cholesterol ratio was unchanged. Additionally, it found that replacing fats with carbohydrates actually *increased* fasting triglyceride concentrations. Overall, the study concluded: "The effects of dietary fats on total:HDL cholesterol [the ratio of total cholesterol to HDL cholesterol] may differ markedly from their effects on LDL." This means that maybe by looking at LDL, researchers had missed the boat—since the total:HDL may have been telling a completely different story.

The latest line of research also shows that the specific type of LDL is important. A rise in the level of the larger, fluffier LDL cholesterol type is now understood to be fine, while a rise in the smaller, denser LDL type is worrisome. The LDL rise associated with saturated fats turns out to be of the lighter, fluffier variety. So even that is of much less concern than once thought.[35]

A 2014 New York Times article, based on an interview with Cambridge University cardiac epidemiologist Dr. Rajiv Chowdhury, notes that "the relationship between saturated fat and LDL is complex." In addition to raising LDL cholesterol, saturated fat also increases HDL, the so-called good cholesterol. "And the LDL that [saturated fat] raises is a subtype of big, fluffy particles that are generally benign," Chowdhury was quoted as saying. The Times article continues, "The smaller, more artery-clogging particles are increased not by saturated fat, but by sugary foods and an excess of carbohydrates, according to Dr. Chowdhury. 'It's the high carbohydrate or sugary diet that should be the focus of dietary guidelines,' he said. 'If anything is driving your low-density lipoproteins in a more adverse way, it's carbohydrates.'"[36]

On top of that, research over the past decade has demonstrated convincingly that having a high HDL number is actually optimal. There is now a broad consensus in the medical community that HDL is actually protective against heart disease. For women, in particular, an elevated HDL number is more predictive that a person will *not* get heart disease than a high LDL number is that she *will*. And fats, including saturated fats, actually raise HDL. In fact, according to Harvard's Dr. Mozaffarian: "The best way to raise HDLs is with saturated fat."[37]

Like the rest of my generation, I grew up hearing the Keys-type saturated fat hypothesis as gospel. So the idea that any kind of fat could be beneficial to cholesterol levels initially struck me as counterintuitive. And it surprised me to learn that highly credible physicians and researchers questioned the Keys hypothesis all along. In fact, though, as the significant rise in heart disease rates in the United States, Britain, and other Westernized countries was getting started, in the 1950s and '60s, competing theories were developed, each with some epidemiological evidence to back it up.

The saturated fat idea was not necessarily ever the theory with the most science behind it. But for various reasons (including, apparently, Keys's forceful personality), it was the idea that won the day, eclipsing other competing theories.

One of the leading opponents of the saturated fat idea, and proponents of an alternative dietary explanation, was the British physician and physiologist mentioned earlier, Dr. John Yudkin. Dr. Yudkin was the highly respected founder of the first university nutrition department in Europe. Like scientists in other parts of the world, he and his colleagues had looked at the available public health data in search of the cause of rising heart disease rates. But unlike Keys, they found the evidence for saturated fat unconvincing.

The More Plausible Explanation: Sugar

Starting in the 1950s, Dr. Yudkin had examined enough epidemiological evidence to believe that rising heart disease rates in Western Europe and the United States were most likely primarily caused by *sugar*, not fat. Unlike Keys, who plotted just 7 countries on a graph, Yudkin looked at the data from 22 countries (including the 7 Keys used) and found a much stronger relationship with sugar. He and his colleagues at the University of London's Queen Elizabeth College soon launched a series of experiments to investigate.

For more than a decade that followed, Yudkin and his research team conducted scores of experiments to determine how sugar behaves in the body. Feeding sugar (often alongside experiments feeding starches or fats) to humans, rodents, and other animals, they closely tracked its effects.[38]

Here's a sampling of some of the team's findings. In rats, their trials consistently found that sugar consumption increased blood pressure, increased blood level triglyceride, and decreased the body's efficiency in dealing with elevated blood glucose. Sugar was also found to make blood platelets "stickier" (more adhesive to one another and to arterial walls), a precursor to atherosclerosis. Other rat experiments, testing levels of two enzymes present in the livers of rats and humans (pyruvate kinase and fatty acid synthetase) greatly increased (fivefold and twofold, respectively), indicating an increased production of fat in the liver. In various rodents, sugar feeding enlarged the liver by 25 to 100 percent. Several experiments showed increased insulin resistance from sugar. In several species (baboons, chickens, pigs, and rabbits), sugar increased cholesterol levels.[39]

Feeding experiments on human subjects (both in Yudkin's lab and elsewhere), subsequently showed similar effects in humans. Sugar in human diets resulted in elevated cholesterol and triglyceride levels. The trials also found a notable rise in human insulin levels: 40 percent after just two weeks on a high-sugar diet. Human adrenal hormones rose significantly, too: by 300 to 400 percent.

As for sugar's mechanism, Dr. Yudkin acknowledged uncertainty. In trying to understand how sugar could be involved with so many diseases, he felt two overriding points were key. "One is that sugar produces an enlargement of the liver and kidneys of our experimental animals, not only by making all the cells swell up a little, but by actually increasing the number of cells in these organs"—conditions referred to as hypertrophy and hyperplasia.[40] The second effect he considered particularly significant was that in some people sugar produced an increase in the levels of insulin, estrogen, and adrenal cortical hormone. Systemic effects like these make it plausible for sugar to be implicated in a large number of conditions. And such effects on the hormones and vital organs, he believed, "should persuade any reasonable person that sugar is not just an ordinary kind of food."[41]

Regarding atherosclerosis, the disease in which plaque builds up on the walls of arteries, blood vessels that carry oxygen-rich blood to the heart and other parts of the body, Yudkin offered a "working hypothesis." His hypothesis was grounded on the belief that its underlying cause is a high level of insulin. Many people known to have hardening arteries also had high blood levels of insulin, he noted. Furthermore, coronary atherosclerosis is accompanied by elevated cholesterol and triglyceride levels, and a range of other disturbances in biochemistry and in platelet behavior. "Only a disturbance of hormone levels is likely to afford an explanation for such a wide variety of changes," he concluded. Various pieces of evidence pointed to insulin as the hormone most likely to be at fault.

Yudkin's investigations were careful, methodical, and empirical. Over the years he published some 300 articles in scholarly journals, including Britain's leading medical journal, *The Lancet*, reporting the results of his research. Having reviewed all extant medical studies on sugar, he remained skeptical about some claims of its harmful effects (including American research connecting sugar consumption to childhood hyperactivity and learning disorders). And he was clear that sugar was not the sole cause of heart disease, which he considered a complex illness triggered by multiple factors.

Ultimately, though, Yudkin found the results from his extensive clinical trials strongly supportive of the epidemiological evidence. Collectively, it was compelling proof that sugar was the primary cause of the modern era's global rise in heart disease.

Sugar affects people differently, he concluded. For nearly all people, sugar fosters tooth decay and gum disease. For many, it contributes to dyspepsia, as sugar can produce or exacerbate an inflamed mucous membrane of the esophagus or stomach. And for some people—not the majority, but a large portion (25 to 30 percent, Yudkin estimated)—whom he deemed sucrose sensitive, sugar triggers a broad array of negative health consequences, including heart disease. Similar to the case of smoking—in which some lifelong smokers get emphysema, some get lung cancer, and some get neither—only some people who eat sugar at levels typical of the modern diet will become afflicted by heart disease.

Throughout the years of his experiments, Yudkin had prolifically presented his findings at nutrition conferences and published articles about the trials. After completing his clinical investigations, Yudkin dedicated a year to summarizing his research and conclusions about sugar in the book *Pure, White and Deadly* (published in 1972, updated in 1986). In it, he pointedly questioned the purported links between heart disease, meat, and saturated fat. Foods, he noted, that humans have been eating for eons.

"Our view, then," he wrote, "is that the underlying cause of coronary disease is a disturbance of hormonal balance. Apart from increased insulin and adrenal hormone, for example, many patients show an increase in estrogen." The agent he believed caused the disturbance was sugar.

Yudkin, as the founder and chair of a leading university nutrition department, was already well known to players in the British and international food industry. The clear condemnations of sugar made in *Pure, White and Deadly* were an unacceptable threat to a burgeoning processed-food industry, which was increasingly devising formulations for foods that could be labeled "low-fat" but were loaded with sugar to replace flavor that had been removed with the fat.

Accordingly, the food industry, joined by rival physician Ancel Keys, deliberately undertook a campaign to discredit Yudkin and his work. "Keys loathed Yudkin and, even before *Pure, White and Deadly* appeared, he published an article describing Yudkin's evidence as 'flimsy indeed,'" according to a biographical newspaper article.[42] Sugar industry trade groups issued press releases dismissing Yudkin's statements as "emotional assertions" and describing his book as "science fiction." When Yudkin sued, they were forced—nearly five years later—to print a retraction. Of course, by that time the damage to his reputation had already been done.

For the rest of his life, Yudkin suffered personally and professionally for his scientific publications and statements about sugar, which the processed-food industry evidently viewed as a serious threat to its bottom line. He found himself "uninvited" to international conferences; forced to cancel conferences he had organized; and blackballed from publishing his findings.[43] His own university even reneged on a promise to allow the professor to use its research facilities in his retirement. "Only after a letter from Yudkin's solicitor was he offered a small room in a separate building."

His despair over the situation is palpable in the words of his updated (1986) version of *Pure, White and Deadly*:

> *Can you wonder that one sometimes becomes quite despondent about whether it is worthwhile trying to do scientific research in matters of health? The results may be of great importance in helping people avoid disease, but you then find that they are being misled by propaganda designed to promote commercial interests in a way that you thought only existed in bad B films.* [44]

As you would expect, other scientists heard and read about Yudkin's work, and some of them took up their own investigations into sugar. Dr. Richard Ahrens, a nutrition professor at University of Maryland who had spent time in Dr. Yudkin's lab, performed U.S. experiments that demonstrated an increase in blood pressure in both rats and humans fed sugar. Dr. Ahrens's experiments and epidemiologic research led him to conclude that coronary heart disease was increasing "on a world-wide scale in rough proportion to the increase of sucrose consumption but not in proportion with saturated fat intake." [45] Similarly, Dr. Sheldon Reiser, at the U.S. Department of Agriculture Nutrition Lab in Beltsville, Maryland, conducted human feeding trials that found about 25 percent of subjects were "sugar sensitive," showing elevated insulin levels with sugar consumption commensurate with the typical American diet. Dr. Reiser's experiments also confirmed that adding sugar to human diets resulted in increased blood levels of triglyceride, cholesterol, and glucose. Even a sugar-industry-funded study (carried out at Wake Forest University) found that pigs fed sugar had elevated cholesterol levels and the genesis of atherosclerosis. [46]

You might also expect the collective weight of such scientific findings to have catalyzed a major push within the medical research community to further investigate sugar's effects on the human body, especially as it relates to coronary disease. Perhaps meat and fats, especially saturated fat, had been wrongly fingered? Instead, Yudkin's line of research virtually dried up and withered away. His public shunning seems to have had a serious chilling effect on other scientists. By the end of the 1970s, a recent biographical article states, "he had been so discredited that few scientists dared publish anything negative about sugar for fear of being similarly attacked."[47]

Part of the problem for Yudkin was that he was so far ahead of the curve. Although his experiments consistently showed detrimental effects from sugar, neither he nor anyone else fully understood the mechanisms. "Three or four of the hormones that would explain his theories had not [yet] been discovered," notes David Gillespie, author of the book *Sweet Poison.*

Epidemiological and clinical research done over the past decade has more than fully restored Yudkin's credibility and breathed new life into his line of inquiry. It has, Dr. Robert Lustig has said, revealed his work as "prophetic." The recent studies go beyond just establishing a link to obesity as a precursor to other diseases. They directly connect sugar consumption to heart disease, diabetes, and other serious illnesses.

Suggesting that sugar makes you fat, which in turn makes you unhealthy (including health problems like diabetes, hypertension, and heart disease), has never been terribly controversial. In fact, other than by Yudkin and a few others, that was the standard assessment of sugar until fairly recently. But the idea that sugar has to make people fat in order to trigger adverse health events has changed. Beyond merely being that empty, excessive calorie, it is now widely recognized that sugar—whether it makes a person fat or not—is bad for a person's health.

Here are a few examples. The Nurses' Health Study, which followed nearly 90,000 women over two decades, found that women who drank more than two servings of sweetened drinks a day had a 40 percent higher risk of heart attacks or death from heart disease than women who rarely consumed sweet drinks.[48]

Similarly, a study that followed 40,000 men for two decades found a 20 percent higher risk of having a heart attack or dying from a heart attack for those who averaged one can of a sugary beverage per day compared with men who rarely or never drank them.[49] Another study found people who regularly consume one or more sugary drinks a day to have a 26 percent greater risk of developing type 2 diabetes.[50] And a 22-year-long study of 80,000 women found that those who consumed a can of sugary drink a day had a 75 percent higher risk of gout.[51]

Recently, such new thinking was exemplified in the February 3, 2014, issue of the *Journal of American Medical Association Internal Medicine*. Along with a major new study linking sugar to death from heart disease, the journal contained a noteworthy commentary by Laura Schmidt, PhD, of the University of California–San Francisco School of Medicine. In "New Unsweetened Truths About Sugar," Dr. Schmidt describes a sea change in perspectives now occurring with respect to sugar. She writes:

We are in the midst of a paradigm shift in research on the health effects of sugar, one fueled by extremely high rates of added sugar overconsumption in the American public. By "added sugar overconsumption," we refer to a total daily consumption of sugars added to products during manufacturing (i.e., not naturally occurring sugars, as in fresh fruit) in excess of dietary limits recommended by expert panels. Past concerns revolved around obesity and dental caries [tooth decay] as the main health hazards. Overconsumption of added sugars has long been associated with an increased risk of cardiovascular disease (CVD). However, under the old paradigm, it was assumed to be a marker for unhealthy diet or obesity. The new paradigm views sugar overconsumption as an independent risk factor in CVD as well as many other chronic diseases, including diabetes mellitus, liver cirrhosis, and dementia—all linked to metabolic perturbations involving dyslipidemia, hypertension, and insulin resistance. The new paradigm hypothesizes that sugar has adverse health

*effects above any purported role as "empty calories" promot-
ing obesity. Too much sugar does not just make us fat; it can
also make us sick.*[52]

The new study accompanying Dr. Schmidt's commentary was
carried out by researchers from Harvard and Emory universities
and the Centers for Disease Control and Prevention. Called "Added
Sugar Intake and Cardiovascular Diseases Mortality Among US
Adults," it connects sugar to heart disease deaths. Noting that epidemiologic studies have long shown an association of sugar with
cardiovascular disease, the researchers sought to address the relative dearth of prospective studies examining sugar's association
with heart disease mortality. They tracked 163,039 "person-years,"
which included 831 heart disease deaths, during a follow-up of
about 15 years.[53] Adjusting for numerous variables, including age,
sex, and socioeconomics, the researchers conclude: "The risk of
CVD [cardiovascular disease] mortality increased exponentially
with increasingly usual percentage calories from added sugar."

More specifically, the study found that the risk of CVD mortality becomes elevated once added sugar makes up more than 15
percent of daily calories. A single can of soda a day increases the
risk of cardiovascular diseases by 30 percent.[54] Noting that "[m]ost
US adults consume more added sugar than is recommended for
a healthy diet," the study concludes: "We observed a significant
relationship between added sugar consumption and increased risk
for CVD mortality."[55]

Increasingly, this is even becoming the message of physicians
communicating to mainstream audiences, like Dr. Mark Hyman,
who writes a health column for *The Huffington Post* and has lately
garnered attention for becoming the health adviser to former president Bill Clinton. "Here's the take-home message: **Fat doesn't make
you fat. Sugar makes you fat**," Hyman has advised. He recommends that
people eat good-quality, "real, whole" foods (including grass-fed
beef), and not worry about fat intake.[56]

Somehow, it's harder to wrap your brain around sugar being
the true culprit for chronic diseases. For one thing, sugar is so

ubiquitous that we tend to assume it must be benign. Fat as a cause of fatness and heart disease also seems logical. We can easily visualize fat globules floating around in our blood and clogging up our arteries. But sugar? How does that work? The idea is much less intuitive.

Despite recent advances in our understanding, the precise mechanism by which sugar works inside the body remains somewhat uncertain. But a great deal of progress has been made since Yudkin's day. Researchers have now established how the body metabolizes the two component carbohydrates of table sugar (sucrose), which are fructose and glucose. It is now understood that while glucose is metabolized by all cells, fructose is primarily metabolized by the liver. "This means consuming excessive fructose puts extra strain on the liver, which then converts fructose to fat." A fatty liver, in turn, can cause metabolic syndrome.

Metabolic syndrome is a relatively new term used to describe a group of risk factors for cardiovascular disease that commonly cluster together. A person is generally considered to have metabolic syndrome when she has three or more of the following five criteria: abdominal obesity, raised triglycerides, reduced HDL, elevated blood pressure, and raised plasma glucose. (Several of these conditions, of course, are the very ones that Yudkin's experiments connected to sugar decades ago.) Most people with metabolic syndrome will also be insulin-resistant (which Yudkin also showed results from sugar consumption); they may or may not have type 2 diabetes (something that various researchers other than Yudkin had contemporaneously connected to sugar).[57] In addition to cardiovascular disease and type 2 diabetes, individuals with metabolic syndrome are susceptible to other conditions, including polycystic ovary syndrome, fatty liver, cholesterol gallstones, asthma, sleep disturbances, and some forms of cancer.[58]

In a 2014 essay titled "Sugar Is Now Enemy Number One in the Western Diet,"[59] London's Croydon University Hospital cardiologist Aseem Malhotra seemed to inherit Dr. Yudkin's mantle, leading the charge to focus the medical and public health communities away from worrying about meat and fats and toward reducing sugar consumption. His article equates the devastating health effects of excessive sugar to smoking, as it compares the food industry's tactics of bullying academic researchers and manipulating the science to those employed by the tobacco industry. Noting that poor diet is actually responsible for *more* disease than smoking, alcohol, and physical inactivity combined, he argues that the bulk of scientific evidence now supports focusing on a particular dietary change: "The evidence suggesting that added sugar should be the target is now overwhelming."

In *Medical News Today*, Malhotra has spoken more specifically about the science behind targeting sugar.[60] Vilification of saturated fat and cholesterol has been misguided, he says. "It is time to bust the myth of the role of saturated fat in heart disease and wind back the harms of dietary advice that has contributed to obesity." Red meat and dairy products, he notes, contain nutrients essential to human health. Additionally, "[r]ecent prospective cohort studies have not supported any significant association between saturated fat intake and cardiovascular risk." To the contrary, Malhotra says, "saturated fat has been found to be protective." Three-quarters of people hospitalized for heart attacks have normal cholesterol levels, he notes.

On the other hand, two-thirds of hospitalized heart attack sufferers have metabolic syndrome.[61] The real issue, Malhotra says, is a related phenomenon, a triad of lipid abnormalities called atherogenic dyslipidemia[62] (or atherogenic lipoprotein phenotype), which is characterized by increased blood concentrations of small, dense-LDL particles, decreased HDL particles, and increased triglycerides. "Atherogenic dyslipidemia is characteristically seen in patients with obesity, the metabolic syndrome, insulin resistance, and type 2 diabetes mellitus."[63] Research done in 2010 by the Cardiovascular Research Center and Center for Human Genetic Research at Massachusetts General Hospital showed that "[d]ietary saturated fat content has little influence on the components of the atherogenic lipoprotein phenotype."[64]

Sugar, by contrast, is directly connected to atherogenic dyslipidemia and metabolic syndrome. Regardless of a person's weight,

Malhotra says, sugar consumption is an independent risk factor for metabolic syndrome. Although the terms for these medical conditions were coined well after Dr. Yudkin wrote *Pure, White and Deadly*, their clusters of characteristics bear a striking resemblance to those he described decades ago in his findings on the effects of dietary sugar.

In fact, these days Yudkin's ideas have gained such widespread acceptance that in March 2014 the World Health Organization lowered its "target" recommendations for added sugars from 10 percent of dietary calories to 5 percent,[65] which is less than the sugar contained in a single can of soda. WHO noted that its recommendation was based on a wealth of studies now connecting sugar to obesity and chronic health problems.[66] Despite moving in this direction, WHO's director-general, Margaret Chan, recently commented on the power of the sugar industry. It seems the sector is not taking such moves lying down, and it will likely continue to fund counter-studies and efforts to discredit scientists doing sugar-damning research.

Of course, the sugar, beverage, and processed-food industries are well aware that Americans now get about 15 percent of their calories from added sweeteners (those that do not occur naturally in foods),[67] and they want to keep it that way. WHO guidelines propose a huge dietary shift. But Dr. Robert Lustig says it's a change that must occur as the cornerstone to addressing the United States' chronic diseases. Lustig has become something of a folk hero for speaking in clear, unvarnished language about the health dangers of sugar despite the aggressive way such scientists have been attacked by industry in the past. Lustig believes that, in addition to obesity and dental problems, sugar is an important factor in heart disease, cancer, Alzheimer's, and diabetes. He also points to science showing that sugar is addictive, like tobacco and cocaine.

Dr. Lustig is among the scientists and physicians who now consider metabolic syndrome the key to the major chronic diseases. "Being thin is not a safeguard against metabolic disease or early death," Lustig writes in his book *Fat Chance*. *"The obesity is not the cause of chronic metabolic disease, otherwise known as metabolic syndrome*. And it's metabolic syndrome that will kill you."[68] Like Yudkin, Malhotra, Schmidt, and others, Lustig does not blame red meat or fat for today's crisis. Sugar, he believes, is the biggest dietary contributor to metabolic syndrome.[69]

And, Lustig argues, there is one simple reason this science has been obscured for so long: the undue influence of the sugar and processed-food industries. Somewhat ironically, his suggestion is only reinforced when you look at Lustig's Wikipedia entry. Nearly two-thirds of the studies cited there to repudiate his views were funded by Coca-Cola.[70]

Having now seen the strong links from sugar to heart disease and other health problems that were once blamed on red meat and fat, the claim that meat and fat are the causes of today's chronic health problems seems even more implausible. But what does the consumption data show about sugar? Let's now return to the federal government's dietary data and take a look.

The upward sugar consumption trend is clear and striking. Between 1900 and 2000, consumption of added sweeteners rose steadily—60 percent over the century and 20 percent between 1970 and 2005. By some estimates, the average American now eats 130 pounds of sugar per year, more than 0.3 pounds a day, every day.[71] That figure tends to shock people (and, in my experience, people immediately aver that their own consumption is much less).

Taking a longer view, though, it's even easier to see that the aberrance of today's sugar consumption is extreme. Humans, Dr. Yudkin pointed out, ate almost no sweetened foods or beverages until quite recently. As he put it, for 99.9 percent of human history, we ate and drank nearly nothing sweet. Sweetness was limited to the fleeting seasons of ripe fruits and the rare treat of wild honey, which is always well guarded by swarms of angry bees.

Sugar as an isolate, an additive, was invented just 2,500 years ago, when a method to extract the sweetness from sugarcane (which, to this day, is no easy task) was devised in India. Still, it remained a rare luxury item for thousands of years.[72] A teaspoon of sugar cost a citizen of Middle Ages Europe as much as a teaspoon of caviar costs us today. For ordinary people, adding a cup of sugar to anything would have been unimaginable. A study on the metabolic effects of sugar in the *American Journal of Clinical Nutrition* notes, "Before the introduction of sugar, the primary sweetener had been honey, but because it was relatively rare and not mass produced, the majority of people (especially the poorer classes) *had no sweeteners at all in their normal diet* even as recently as the early 19th century."[73]

By 1800, per-capita annual American sugar consumption is estimated to have risen to around 6 to 18 pounds (much more than in ancient times, but, even at the higher end, only around 15 percent of today's level). Total global sugar production amounted to just 0.25 million tons.[74] By the early 20th century, sugar was quite common, but it was still expensive compared with today's prices. In 1920, a 5-pound bag of sugar cost the equivalent of $7.61 (almost three times more than today).[75]

A graph of U.S. sugar consumption since 1800 is pretty much a line heading upward at about 45 degrees. The years during World War II were the exception. Sugar was rationed by the U.S. government from 1943 to 1946,[76] and graphs of U.S. sugar supply dip during those years.[77] Beef, cheese, and other meats were also rationed. Epidemiological studies have sometimes pointed to a subsequent dip in heart disease as a result of rationing those animal-based foods. But isn't it just as likely—in fact, much *more* likely given what we now know—that sugar rationing is the real explanation?

These days, the Food and Agriculture Organization says that by weight sugarcane is the world's largest crop. In 2012, it was cultivated on 31 million hectares (26 million in cane, 5 million in beet), producing 173 million tons of sugar.[78] As mentioned previously, U.S. consumption and that for many other countries is estimated to top 100 pounds per year. In other words, our sugar consumption has exceeded the levels our bodies evolved with for more than two centuries, and has increased *ninefold* over the past 200 years. It is not hyperbole to label today's levels, as the above-referenced journal study did, "an epidemic of sugar consumption."[79]

In today's diet, a lot of the sweetness comes from soda and juicy drinks. From 1977 to 2001, sweetened beverage consumption more than doubled, accounting for 278 additional calories in the daily diet.[80] The average American now drinks over 50 gallons of soda a year.[81] The USC Childhood Obesity Research Center at the Keck School of Medicine calls the jump in soda consumption "the prime driver behind the obesity epidemic."[82] Research specifically links sweet beverages to heart disease and adverse changes in lipids, inflammatory factors, and leptin (the hormone that controls weight gain).[83]

Some research suggests that the shift to high-fructose corn syrup as a sweetener in beverages and other processed foods has further exacerbated our current health crisis.[84] (Beet and cane sugar consumption have actually slightly declined in the past three decades as consumption of HFCS has risen a staggering 10,000 percent.) Sucrose, remember, is half fructose, and fructose is now believed to be the cause of most of sugar's negative health consequences.

Researchers at the University of Southern California have recently discovered that the high-fructose corn syrup in many sodas is as much as 65 percent fructose, nearly 20 percent higher than commonly assumed. "The elevated fructose levels in the sodas most Americans drink are of particular concern because of the negative effects fructose has on the body," explained study author Michael Goran, PhD, professor in the departments of preventive medicine, physiology and biophysics, and pediatrics at the Keck School. "Unlike glucose (the smaller component of HFCS), over-consumption of fructose is directly responsible for a broad spectrum of negative health effects." The body processes fructose differently from glucose, Goran points out. Consuming large amounts of fructose (other than when naturally occurring in whole fruits) greatly exacerbates the risk for those diseases by also causing fatty liver disease, insulin resistance, increased triglyceride levels, and an acute rise in blood pressure—in other words, by bringing about metabolic syndrome.[85]

As I've studied the data on rising sugar intake, and the emerging science about the toll it takes on our bodies, again and again, my thoughts turn to my father. At 42, his parents, within a few months of each other, both died of heart failure. They were in their early 70s. Shaken, my father was determined to have a different fate. Convinced their diets were largely to blame, he immediately swore off sugar (which my father considered partly to blame for his mother's obesity and diabetes and which, apparently, they had both consumed in abundance). A man of extraordinary self-discipline, he refused sweets for the next 40 years. With the exception of a small slice of home-baked cake on his birthday and a rare piece of homemade fruit pie, the only sweet things he consumed were whole fruits and a small glass of orange juice with his breakfast.

Sometime in high school, I began gently needling my father about his diet, which seemed a bit overboard. Candy was my vice and I always made sure to have a little stash in my backpack (the junkier, the better—Everlasting Gobstoppers and Lemonheads were favorites). Yet I considered my sweet tooth defensible because I was skinny and very physically active. Taking the conventional view at the time, I'd tell my father: "The only thing wrong with sugar is that it doesn't have any nutrients." I remember using a phrase I'd learned from Jane Brody's *New York Times* health column, "It's just empty calories." And so I (and countless dietitians) believed for many years following. My father was very much a traditionalist, but, on this issue, he was way ahead of his time. He died several years ago, at age 85, from causes other than heart disease. I now believe his choice to avoid sugar added a decade to his life.

Yudkin, Lustig, Malhotra, and Schmidt are just the tip of the iceberg of changing views about sugar. Physicians and scientists are now beginning to agree that epidemiologic and laboratory research, as well as clinical trials, clearly show that sugar is not the benign substance it was once considered. And no one, not even the young, fit, and trim, is immune to its effects.

Beyond risks to coronary health and the connections to diabetes and metabolic syndrome, there is also evidence that sugar is connected with loss of brain function. Shaheen E. Lakhan, MD, PhD, MEd, executive director of the Global Neuroscience Initiative Foundation, believes "non-natural fructose" affects mental health issues, particularly cognitive decline and memory.[86] "[S]everal studies have shown that intake of added sugars is associated with lower cognitive function . . . [T]he association between added sugar intake and MMSE [metabolic syndrome] was independent of BMI [body mass index] and age." A recent *Medscape* article noted that Dr. Lakhan has been at the forefront of some of the newest research on mechanisms connecting diet and mental health, specifically suggesting a relationship with the microbiota in the gut. "The emerging evidence suggests the pathway to be diet-microbiota-inflammation-mental health," he explains.[87] Although this science is in its infancy, it suggests how our gut flora, which are among the first cells to come in contact with our food, may play a role in these adverse outcomes.

Dealing with the scourge of dietary sugar is an enormous public health and medical challenge. Humans instinctively seek out sweet foods. Exceptionally rare in nature, sweetness signals something ripe and good. It is said there is no known naturally sweet food that is poisonous. Fruits are ripe only fleetingly, and typically come around just once a year. And the sugar in fruits is tightly bound to loads of fiber, making it slow to digest and nearly impossible to overeat. Most experts agree sugar that occurs naturally in fruits and vegetables, including naturally occurring fructose, is not the problem.

Compounding the public health challenge, physiologically sugar functions like an addictive substance. Recent studies have documented that sugar affects the body and brain in much the same manner as drugs like cocaine.[88] And like other drugs, it fosters a nearly irrepressible urge to consume more of it. Research at Princeton found that rats can be turned into sugar addicts.[89] Clinical research on humans, done by the Oregon Research Institute in 2013, demonstrated that the brain's "reward center" is activated by sweetness much more than fat.[90] It's the sugar we crave.

Carbohydrates, particularly sugar, are uniquely effective at triggering the brain's reward center. Beyond an aesthetic state, "happiness is also a biochemical state, mediated by the neurotransmitter serotonin," explains Robert Lustig. "One way to increase serotonin synthesis in the brain is to eat lots of carbohydrates." Yet, analogous to other addictive substances, over time our bodies require more and more of them to get the same sensation. "Eating more carbohydrates, especially sugar, initially does double duty: it facilitates serotonin transport and it substitutes pleasure for happiness in the short term. But as the D2 receptor [in the brain] down-regulates, more sugar is needed for the same effect."[91]

Here I'll pause momentarily for a personal confession: I have had a lifelong sugar habit. Research about the addictive nature of sugar completely aligns with my own experiences. I spent a decade working to curb my longings for sugar—eliminating all sweetened beverages, cutting out candy almost entirely, and always taking mouse-sized portions of any sweet snack or dessert. Nonetheless, I continued to have an almost uncontrollable urge to eat something sweet every day after lunch and dinner. Like Pavlov's dogs hearing that bell: *Finish meal, crave sweets.* The only way I have found to rid myself of the longings for sugar is to completely eliminate it from my diet. Research now suggests my lifelong sweet tooth is both learned and biological.

Much as I adore and appreciate her, my mom may bear some of the blame. Like clockwork, she's always had a little sweet thing every afternoon and every evening. Cutting-edge research suggests that a mother's diet affects her baby's food preferences as it develops in the womb.[92] Perhaps some part of my sweet craving was already in place by the time I was born. As Americans have been eating more and more sugar, developing fetuses have also been getting exposed to ever-increasing amounts. It makes sense that the American sweet tooth has been a self-perpetuating and intensifying cycle.

On the plus side, while standard medical advice of the day was, "Why bother with breast-feeding?" my mom did. (She was very thrifty, and I suspect this was largely an economic decision on her part—breast milk is free, and formula costs a small fortune— but no matter.) A study on formula feeding in the United States notes: "During the 1950s and 1960s, the trend in breast-feeding was steadily downward, and by the early 1970s, only 25 percent of infants were breast-fed at age one-week and only 14 percent between two and three months of age."[93] In his book *Fat Chance,* Dr. Robert Lustig explicitly mentions "less breastfeeding" as among the major causes of today's obesity epidemic.[94] America's switch to feeding babies with formula is almost certainly another factor in today's chronic health problems.

The cloying sweetness of formula is part of the problem. Human breast milk contains no sucrose at all. Sucrose, remember, is a combination of fructose and glucose. Breast milk contains only lactose (7.2 percent), which is considerably less sweet.[95] In the 1950s, sucrose was the primary sweetener in baby formula, and, while it has been taken out of some formulas, it remains in major brands on U.S. grocery shelves. Similac Organic and Similac Soy Formula each contains about 1 teaspoon of sucrose in every 5 ounces. A *New York Times* blind tasting panel of infant formulas unanimously identified those sweetened with sucrose as far sweeter than those with lactose.

Overly sweet baby formula contributes to two sets of problems. For one, it triggers overeating. Dr. Benjamin Caballero, director of the Center for Human Nutrition at the Johns Hopkins Bloomberg School of Public Health and an expert in risk factors for childhood obesity, is very concerned about unnaturally sweet infant formulas. "[S]weet tastes tend to encourage consumption of excessive amounts," Dr. Caballero says. He points to evidence showing that babies and children will always show a preference for the sweetest food available. They will eat more of it than they would of less-sweet food.[96]

On top of that, highly sweet formulas may start babies down the path of lifelong sugar craving. Pediatric dentist and nutritionist Dr. Kevin Boyd says of sucrose-sweetened formulas: "We're conditioning them to crave sweetness." Formula with sucrose, which he calls "super sweet," will foster a cycle of children craving sugar, and "makes the kid want to eat more."[97] According to Dr. Ivan de Araujo, a fellow at the John B. Pierce Laboratory at Yale University School of Medicine, new research shows that animals prefer sucrose over other sugars and that eating sucrose generates sugar cravings. He agrees that putting sucrose in baby formulas could have negative long-term effects.[98]

Although it's uncertain exactly why (how much is formula composition, how much is overconsumption, how much is it the bottle as delivery mechanism?), a long string of studies have demonstrated that formula-fed babies are more likely to end up with various health problems, including being overweight. A 2013 study by researchers at Brigham Young University found that compared with breast-fed babies, infants fed formula in the first six months were more than twice as likely to be obese by age two.[99] Studies have also connected formula feeding with a wide range of baby health problems, including ear infections, food allergies, and asthma.[100] A 2012 study published in the journal *Pediatrics* estimated that if 90 percent of U.S. families followed guidelines to breast-feed exclusively for six months, the country would annually save $13 billion from reduced medical and other costs.[101]

Because of findings like these, the American Academy of Pediatrics recommends exclusively breast-feeding babies until the age of six months, and continuing to breast-feed after the introduction of solid foods until the baby is at least one year old. In reality, though, only 20 percent of U.S. infants are nursed at all by the time they reach six months.[102] Bottle-feeding will continue to aggravate the obesity crisis in the years ahead.

Alternative Hypothesis, Part 2: Grains

While the case against sugar is very nearly undeniable, there is another group of carbohydrates that may also be contributing to the rise in obesity and chronic diseases: grains.

In contrast with meat and animal fats—or even sugar—the safety and healthfulness of grains have rarely been questioned.

Expressions like *the bread of life* indicate the central role grains have held in the diets of some humans in recent millennia. Today, wheat comprises 20 percent of the world's dietary calories. But some physicians and scientists have begun to question whether grain-based foods—especially from modern strains—should be a dietary staple.

This is largely because the body treats all refined carbohydrates in a similar fashion. Glycemic index is a measure of how foods affect blood sugar and insulin. The higher a food's glycemic index, the more it affects blood sugar levels. "Foods with a high glycemic index, like white bread, are rapidly digested and cause substantial fluctuations in blood sugar," notes the Harvard School of Public Health's newsletter.[103]

Yet Dr. William Davis, a preventive cardiologist who practices in Milwaukee, Wisconsin, argues in his book *Wheat Belly* that the problem goes well beyond white flour. Bread, he believes, is not the staff of life but a danger to human health.[104] Davis considers much of the problem to originate from the modern, high-yield dwarf wheat plant, a faster-growing plant that's about 2.5 feet shorter than older varieties of wheat with a much larger seed. Unfortunately, even whole wheat is problematic. The human body rapidly converts modern wheat to sugar because it contains amylopectin A, a carbohydrate unique to wheat, Davis says, adding that eating two slices of whole wheat bread increases your blood sugar more than a candy bar does. Blood sugar soon plunges and you feel hungry again, leading to weight gain and, specifically, the accumulation of visceral fat (fat around your internal organs).

Dr. Davis's studies and personal observations have led him to become concerned about possible health risks from wheat products. He started telling his pre-diabetic and diabetic patients to remove all wheat from their diets. Dramatic health improvements were realized by those patients who have taken his advice, Davis says. Eliminating wheat from the diet has led his patients to weight loss as well as improvements in asthma, acid reflux, mental clarity, and more.[105] "I've seen this with thousands of patients."

Moreover, Dr. Davis argues that carbohydrates pose the greatest risk for formation of small low-density lipoprotein [LDL] particles (remember—those are the bad ones), leading to athero-sclerotic plaque, which in turn triggers heart disease and stroke. Eating carbohydrates increases heart disease risk, he says, "even if you're a slender, vigorous, healthy person."

Additionally, "carbohydrates increase your blood sugars, which cause this process of glycation, that is, the glucose modification of proteins," Davis says. The glycation Davis speaks of is the bonding of a protein or lipid molecule to a sugar molecule, fructose or glucose. This process creates inflammation, which can activate your immune system.[106] And glycating small LDL makes one more prone to atherosclerosis.

Davis's skepticism of carbohydrate-heavy diets, and especially wheat, is shared by Dr. David Perlmutter, a board-certified neu-rologist and a fellow of the American College of Nutrition who practices in Naples, Florida. "Carbohydrates are absolutely at the cornerstone of all of our major degenerative conditions," Perlmutter states unequivocally.[107] He includes Alzheimer's, heart disease, and cancers. Elevations in blood sugar, even mild ones, compromise brain structure, lead to brain shrinkage, and "are strongly related to developing Alzheimer's disease," according to Perlmutter. Grain foods have a high glycemic index, cause carbohydrate surge and inflammation, and are therefore detrimental to the brain, he believes. "Inflammation is the cornerstone of Alzheimer's disease and Parkinson's, multiple sclerosis—all of the neurodegenerative diseases are really predicated on inflammation."

The human diet consists of three macronutrients: carbohy-drates, protein, and fat. Americans today eat about 60 percent carbohydrate, 20 percent protein, and 20 percent fat. These levels contrast starkly with the optimal diet, according to Perlmutter, which is 75 percent fat, 20 percent protein, and just 5 percent carbohydrates.

To those who accuse Perlmutter of advocating extreme dietary change, he answers that what he is suggesting is not change. Rather, it is getting back to a diet our bodies evolved with. "I am recommending that we end this grand experiment and return to a diet that isn't evolutionarily discordant." The extreme change is the shift in the human diet in recent centuries, he notes. "In the early 19th century, Americans consumed just over 6 pounds of sugar each year. That figure now exceeds 100 pounds. And there has been a dramatic reduction in the consumption of healthful fat." The consequences of the modern diet are protein glycation, uncontrolled insulin signaling, and unknown epigenetic consequences.

In addition to inappropriate diets, Perlmutter considers widespread vitamin D shortages another reason for today's epidemic of chronic diseases. "Vitamin D is a powerful player in terms of brain health," he says. "Beyond strong and healthy bones, vitamin D activates more than 900 genes in human physiology, most of which are important for brain health. Low levels of vitamin D correlate with increased risk for multiple sclerosis, dementia, and Parkinson's disease." As I suggested earlier, the disappearance of independent farms and ranches is one of the many societal changes exacerbating our population's vitamin D shortage.

The Carbohydrate Connection

Gary Taubes, author of the best-selling books *Good Calorie, Bad Calorie* and *Why We Get Fat*, argues that the problem in the modern diet is excessive carbohydrates—too much "sugar and flour," as he refers to it in shorthand. Metabolizing carbohydrates requires the body to release insulin, which in the short term leads to weight gain, and in the long term leads to insulin resistance, Taubes argues.

Part of what's intriguing about Taubes's argument is the demographic data for carbohydrates. We've already seen that sugar consumption has skyrocketed. The same thing has happened with grains over the past three decades. From 1970 to 2000, per-capita U.S. grain consumption of grains increased by an astonishing 48 percent.[108]

Aside from heart disease, other chronic diseases that are also regularly pinned on red meat include obesity, diabetes, and sometimes even brain degeneration. For each of these, as previously explained, sugar is now a considerably more credible suspect. And ironically, there is also considerable evidence of just the reverse: that red meat may help avoid those problems.

In his books, Gary Taubes traces more than a century of clinical treatments for obesity and diabetes that found a single strategy most successful: limiting carbohydrates. Not only were beef and beef fat allowed, but diets like Dr. Atkins's low-carbohydrate one specifically allowed both without restriction. Until a decade or so ago, such diets were controversial because they ran so directly counter to the prevailing high-carbohydrate, low-fat diet that had so thoroughly captivated America's public health and medical communities.

"Over the past five years, however," Taubes wrote in 2002, "there has been a subtle shift in the scientific consensus. It used to be that even considering the possibility of the alternative hypothesis, let alone researching it, was tantamount to quackery by association. Now a small but growing minority of establishment researchers have come to take seriously what the low-carb-diet doctors have been saying all along."[109]

Even conservative institutions like the Harvard School of Public Health have acknowledged the shift. Posing the question in its January 2014 newsletter: "Will going on a low-carbohydrate diet help me lose weight?" it acknowledged "much debate about the impact of low carbohydrate eating pattern on overall health." It then cited a recent 20-year study on 82,080 women, which "found that a low-carbohydrate eating pattern did not increase risk of heart disease." Harvard School of Public Health is now doing its own controlled trial, the newsletter stated, which it believes will provide "a more definitive answer on the possible benefits of low-carbohydrate diets."[110]

New research seems to be shifting the consensus. Just as many studies now document damage to the human body from excessive consumption of sugar and other carbohydrates, a growing body of research is also documenting the efficacy of low-carbohydrate diets for weight loss and diabetes control.

One study has been especially influential. A well-funded yearlong clinical trial carried out by researchers at the Prevention Research Center at the Stanford University School of Medicine was billed as "the largest- and longest-ever comparison of four popular diets."[111] Previous clinical trials, the researchers noted, had found that low-carbohydrate, non-energy-restricted diets "were at least as effective as low-fat, high-carbohydrate diets in inducing weight loss." Yet they were limited by small sample sizes, high rates of attrition, short durations, or limited diet assessment. For the Stanford study, more than 300 women were enrolled, each randomly assigned to one of four groups, with diets ranging from very-high-carbohydrate, low-fat diets (the "Ornish" group), to a low-carbohydrate, high-fat diet (the "Atkins" group). The other two diets had intermediate levels of fats and carbohydrates.

The title of the press release suggests the study's unequivocal findings: "Stanford Diet Study Tips Scale in Favor of Atkins Low Carb." "The lowest-carbohydrate Atkins diet came out on top," the press release began, continuing: "Those randomly assigned to follow the Atkins diet for a year not only lost more weight than the other participants, but also experienced the most benefits in terms of cholesterol and blood pressure."

It's important to note that the Stanford study went beyond establishing the efficacy of low-carbohydrate diets for weight loss: It also exonerated them from suspicions that they are otherwise unhealthy. In an article published in the *Journal of the American Medical Association*, the researchers concluded: "We interpret these findings to suggest that there were no adverse effects on the lipid variables for women following the Atkins diet compared with the other diets and, furthermore, no adverse effects were observed on any weight-related variable measured in this study at any time point for the Atkins group."

Unlike the all-too-common scenario where experiments are carried out to "prove" what researchers want to show, the Stanford team had not initiated the study to advance low-carbohydrate diets. Quite the reverse. The lead researcher, Christopher Gardner, PhD, a longtime vegetarian, acknowledged having launched the study partly due to his concerns that Atkins-type diets might be dangerous. "Many health professionals, including us, have either dismissed the value of very-low-carbohydrate diets for weight loss or been very skeptical of them," he said. "But it seems to be a viable alternative for dieters."[112]

Concerns over negative health effects of low-carbohydrate diets, high in total and saturated fat, simply did not pan out, the researchers reported. "[R]ecent trials, like the current study, have consistently reported that triglycerides, HDL-C, blood pressure, and measures of insulin resistance either were not significantly different or were more favorable for the very-low-carbohydrate groups."

The only exception seemed to be LDL cholesterol, which went up for the low-carbohydrate dieters. But even here, the researchers concluded the effect was not a concern. "Although a higher LDL-C concentration would appear to be an adverse effect, this may not be the case under these study conditions." They went on to explain that the triglyceride-lowering effect of a low-carbohydrate diet leads to an increase in LDL particle size, which is known to decrease LDL atherogenicity. While the LDL concentrations increased slightly (2 percent), triglyceride concentrations decreased by 30 percent. The study did not assess LDL particle size, but the researchers concluded that their findings were consistent with a beneficial increase in LDL particle size. Thus, the Stanford researchers concluded, the low-carbohydrate diet was the most effective for weight loss and had no noteworthy downsides.

What About Diabetes?

As mentioned previously, a considerable amount of research links type 2 diabetes to sugar, both directly, through the ways it affects the body's metabolism, and indirectly, by contributing to obesity. Some studies have also purported to show a link between diabetes and red meat. However, there are serious problems with this idea. One is the healthy user bias discussed previously: People who eat less red meat and saturated fat also tend to eat less refined sugar, trans fats, processed foods, and the like; they are also more physically active, and don't smoke. People who are highly motivated by health and fitness concerns have been steadily fed a diet of anti-beef propaganda for decades, and that has undoubtedly had a significant effect on just who are the regular consumers of beef. People who are more likely to develop diabetes, in other words, may well be the same people who would likely ignore or be unaware of the advice to avoid red meat.

Additionally, as the Harvard study referenced above notes, research generally fails to separate processed and unprocessed meat, but those studies that do have found a much stronger association with negative health effects—including diabetes—from processed meats. This suggests that something (salt, nitrites, or the like) other than the meat itself is responsible for the observed relationship.[113]

The most important study to date that seems to support a connection between diabetes and red meat is one carried out by a group of researchers from various institutions, including Harvard's School of Public Health, in 2013, and it provided only weak evidence of the link.[114] Like many of the studies that have claimed a link to heart disease, this was an observational study in which the researchers examined data, rather than a clinical study. For reasons already discussed, this places limits on the weight it should be given. In addition, for this particular study, the results were far from conclusive.

As explained by Chris Kresser in his article "Does Eating Red Meat Increase the Risk of Diabetes," all but one of the associations between red meat and diabetes disappeared after the researchers adjusted for body mass index (BMI)—in other words, body weight. "[E]xcess body fat is the biggest risk factor for type 2 diabetes, so it shouldn't come as a surprise that people with higher BMIs have less-than-ideal biomarkers for glucose metabolism," Kresser points out about this study. "Additionally, it's common for people who are overweight or obese to have underlying chronic inflammation, so it makes sense that people with higher BMIs would tend towards having higher levels of inflammatory biomarkers."[115]

Mark Sisson, "primal diet" advocate and author, notes that this study was heavily impacted by the healthy user bias. Not all important dietary and lifestyle factors were given consideration by the researchers (sugar consumption levels, for example), but even the available data shows clearly that the "heavy meat eaters" of the study had much less health-conscious diets and lifestyles.

> Folks in the highest quintiles of meat intake were the least active and the most sedentary. They exercised the least and smoked the most tobacco. They drank more alcohol than any other quintile. They guzzled more soda and other sweetened beverages. In the high meat quintiles, folks ate 800 more calories per day than folks in the low meat quintiles. They were much heavier, too . . . Trans fat intake was higher in the high-meat quintiles, too, as was potato intake . . . They ate the least amount of fiber from grains, indicating they probably ate the most refined grains, drank the most coffee, and ate the fewest fruits and vegetables. In short, people who ate the most red and processed meat were also the unhealthiest by both Primal and mainstream standards . . . These people (all health professionals, ironically) most likely didn't particularly care about their health . . . Added sugar . . . wasn't [even] covered.[116]

Moreover, as Taubes argues compellingly in his aforementioned books on weight loss and obesity, the greatest cause of obesity today is excessive carbohydrate consumption. Beef, rich in protein and healthy fats, is a valuable part of a diet intended to take and keep off excess body weight. The Stanford study demonstrated both the efficacy and safety of the Atkins diet, which allows dieters to eat as much beef as they choose.

A Novel Theory: TMAO

Recently a new theory related to red meat consumption has surfaced, this time to do with the way red meat affects the microbes that live in our digestive tracts. Researchers from the Cleveland Clinic in 2013 hypothesized that eating red meat predisposes the gut flora toward production of trimethylamine N-oxide (TMAO), which has been associated with increased heart attack risk.[117] The clinic's news release also stated that the same level of TMAO production did not occur in the body of a vegan, even when she ate meat. The red meat–TMAO theory is a novel idea with very little science to back it up. Commenting on the Cleveland Clinic study, Harvard cardiologist Dr. Dariush Mozaffarian stated it was simply "too early" to know whether the TMAO molecule causes atherosclerosis in humans.[118] Moreover, recent studies have shown that excluding meat from your diet does not lead to greater longevity or better overall health.[119] The hypothesis also fails to explain why major epidemiological studies, like the ones I've cited previously, have shown no relationship between red meat consumption and heart disease.

Moreover, other differences in diet or behavior may actually explain the higher presence of TMAO in the blood of meat eaters. For example, the red meat eaters involved in this study may also have eaten fewer fruits and vegetables and more refined carbohydrates, flour, and sugar. Numerous previous studies have shown such eating habits to be more prevalent among red meat eaters, and all of them have been associated with undesirable changes in the gut microbiota.[120] Previous work by the same researchers showed that people with higher levels of certain gut bacteria produce higher levels of TMAO; research by others has shown that the same gut bacteria is associated with consumption of whole grains.[121] The TMAO hypothesis is a long way from being a credible indictment of beef consumption.

Is Red Meat Linked with Cancer?

Issues similar to those that have made heart disease research unreliable have clouded the discussion over red meat and cancer. Research into the question is fraught with problems. For one thing, "Despite numerous investigations, to date none of these hypotheses have been able to convincingly explain the link between red meat intake and cancer risk," notes a 2013 *Medscape* article.[122] As with heart disease and diabetes, studies have undoubtedly been heavily impacted by the healthy user bias discussed previously. This bias may have an even greater effect on cancer studies. This is because fruits and vegetables seem to confer an anti-cancer benefit, and many of the most health-conscious people not only avoid red meat but also consume fruits and vegetables in abundance. So the very segment of the population eating the most fruits and vegetables overlaps considerably with the population eating little or no beef. This renders conclusions about beef's connection to cancer much more suspect.

Additionally, a number of studies have failed to make crucial distinctions in the data. Some have failed to consider the form of the meats—fresh versus processed. Other studies have failed to consider the means of preparation—grilled or braised. As explained previously, these differences are essential because there is strong evidence that both factors play an important role in the healthfulness of meat when it is consumed. For instance, a major European study in 2013 in which a large cohort (almost half a million individuals) was followed for a median of 12.7 years found a higher cancer risk for processed meats (like sausages and bacon) but no increased risk for unprocessed meats.[123]

In fact, the association between unprocessed red meats and cancer is, at best, weak. The type of cancer given the most media attention on this question has been colon cancer. To wit, a 2011 study in the *European Journal of Cancer Prevention*, "Meta-Analysis of Prospective Studies of Red Meat Consumption and Colorectal Cancer," sought to address the considerable "scientific debate" about the association between red meat and colorectal cancer (CRC). The authors conducted a meta-analysis of 25 independent, non-overlapping prospective studies. Researchers found the associations "weak" and "inconsistent," and found "the likely influence of confounding by other dietary and lifestyle factors." They conclude: "The available epidemiologic data are not sufficient to support an independent and unequivocal positive association between red meat intake and CRC."[124]

Other than colon cancer, the possible association between red meat and breast cancer seems to have garnered the most attention. Yet a major study by researchers from the Harvard School of Public Health in 2002 found no credible link between breast cancer and milk or meat. Researchers sought to assess "the risk of breast cancer associated with meat and dairy food consumption and . . . whether non-dietary risk factors modify the relation." More than 20 studies, they noted, had previously investigated the issue with conflicting results. The Harvard team combined the primary data from eight prospective cohort studies from North America and Western Europe with at least 200 incident breast cancer cases, assessment of usual food and nutrient intakes, and a validation study of the dietary assessment instrument. The data included 351,041 women, 7,379 of whom were diagnosed with invasive breast cancer during up to 15 years of follow-up. The team's investigation reached the following conclusion: "We found no significant associations between intake of meat or dairy products and risk of breast cancer."[125]

Other recent studies have reached similar conclusions for ovarian cancer[126] and pancreatic cancer.[127]

Clearly, multiple, diverse factors have conspired to create our modern epidemic of chronic diseases. Some of these—such as the overuse of antibiotics, and the ubiquity of genetically modified crops like soy and corn—are just coming under scrutiny as possible contributing factors. Between 85 and 95 percent of several major crops are now genetically modified. Some of these are foods eaten directly by humans—like papayas, sugar beets, and canola. Some, like corn and soy, are fed to animals. The long-term health and environmental consequences of this are both unknown and unknowable.

But one thing is patently clear: The surge in chronic diseases is not due to beef or animal fat, which we eat considerably less of now than we did at the dawn of the 20th century. As Harvard's Dr.

Mozaffarian has explained in reference to heart disease in the talk, "What Is the Optimal Diet to Prevent Cardiovascular Disease?" neither saturated fat nor red meat is much of a risk factor.[128]

Rather, as Mozaffarian also notes, the evidence now strongly points to processed foods, especially those containing sugar and white flour, as the major dietary contributors to today's chronic diseases. Sugar is uniquely problematic because of its addictive nature. Eating sugar makes us crave sugar. We're getting it in the womb. It's fed to us as soon as we're born. Sweet foods and drinks in mothers' diets, formula feeding, followed by a diet loaded with daily doses of sugar have likely created whole generations with more intense sweet cravings and have definitely resulted in a diet more sugar-laden than our forebears'. And unlike red meat and animal fats, our consumption rates for sweeteners have increased in parallel with the rising rates of chronic diseases. It is increasingly clear that sugar—not red meat or animal fat—is the primary reason for today's high rate of chronic diseases.

Despite the undeniable fact that worsening U.S. health has *not* been accompanied by rising consumption rates of red meat and animal fat (the reverse is true, actually), the notion that those foods are linked to obesity, diabetes, and heart disease dies hard. The idea is still widely accepted not only in the general public but in the public health and medical community as well. As a lawyer and biologist, I know there's nearly always more than one way to prove or disprove a hypothesis. Moving beyond the demographic data, then, let's approach this question another way.

Health and Diets of Hunter-Gatherers

We can test how red meat and animal fat affect human health by examining the physical condition of people who eat copious amounts of them. If those foods cause heart disease and other health problems, those illnesses should be especially prevalent among those groups. It's a simple test of cause and effect. If factors other than animal fats and red meat are at work, then the association would likely be absent.

Additionally, since I believe the true culprits for the poor state of modern health are sugar and flour, ideally we want to look at populations that do *not* eat those foods. Unfortunately, sugar and flour have been widely distributed throughout the globe for a long time, making this difficult. But it's not impossible.

Hunter-gatherer societies rely heavily on animal-based foods. They cultivate neither grains nor sugar; up until the 20th century, many also did not have access to those foods through trade. Anthropology professors Michael Gurven and Hillard Kaplan examined hunter-gatherer health,[129] including chronic diseases and longevity. They concluded that while infant mortality and death by accident and injury were common, chronic diseases are virtually unknown. On average, chronic diseases accounted for just 9 percent of all deaths. They write:

> *Degenerative deaths are relatively few, confined largely to problems early in infancy and late-age cerebrovascular problems, as well as attributions of "old age" in the absence of obvious symptoms or pathology. Heart attacks and strokes appear rare and do not account for these old-age deaths, which tend to occur when sleeping. . . . Obesity is rare, hypertension is low, cholesterol and triglyceride levels are low, and maximal oxygen uptake is high. Overall, degenerative disease accounts for 6–24 percent (average 9 percent) of deaths . . .*[130]

In his compendium *On Food and Cooking*, Harold McGee states that "[t]hanks to the combination of meat, calcium-rich leaf foods, and a vigorous life, the early hunter-gatherers were robust, with strong skeletons, jaws, and teeth."[131] With settlement and the development of agriculture, human health steadily declined. But late in the 19th century, there was a "return to something like the robustness of the hunter-gatherers," according to McGee. "[T]he growing nutritional contribution of meat and milk . . . played an essential role."[132]

Sometimes it is suggested that chronic diseases fail to manifest in hunter-gatherer communities because people simply die too young. Some earlier researchers had claimed that hunter-gatherers had lives that were "nasty, brutish, and short," rarely living past age 40 or 50. However, Gurven and Kaplan's study found no truth in that. Instead, for individuals who make it to maturity, it was not uncommon for them to reach 68 to 78 years of age.

Of particular value in this inquiry are analyses of historical hunter-gatherer groups who ate meat and animal fat, but did not eat sugar and flour. At least three such populations exist: the Inuit, other Native Americans, and the Masai.[133]

Native American diets varied by climate and locale but, particularly before the start of agriculture, were all heavily based on the flesh and fat of animals. Among them were "deer, buffalo, wild sheep, goat, antelope, moose, elk, caribou, bear and peccary" as well as small animals, including "beaver, rabbit, squirrel, skunk, muskrat and raccoon; reptiles including snakes, lizards, turtles, and alligators; fish and shellfish; wild birds including ducks and geese; and . . . sea mammals."

People commonly assume the wild animals consumed by Native Americans were lean, and that, correspondingly, Indians consumed little animal fat, and negligible quantities of saturated fat. Zoological information about the animals they ate strongly suggests otherwise. Other than seals and squirrels, "the fat of all the other animals that the Indians hunted and ate contained less than 10 percent polyunsaturated fatty acids, some less than 2 percent. Most prized was the internal kidney fat of ruminant animals, which can be as high as 65 percent saturated."[134]

Additionally, historic accounts demonstrate the value traditional Native Americans placed on animal fats. Anthropologist and Arctic explorer Vilhjalmur Stefansson, who reported in detail about the years he spent living among with northern Native Americans, noted that they preferred "the flesh of older animals" rather than that of calves, yearlings, or two-year-olds. Older animals were prized for their thick slab of fat along the back. "In an animal of 1,000 pounds, this slab could weigh 40 to 50 pounds. Another 20 to 30 pounds of highly saturated fat could be removed from the cavity." All of this fat was saved, some rendered, and used to flavor and preserve foods. Animal fat contributed as much as 80 percent of total calories in the diets of the northern Indians.[135]

The traditional diet of the Inuit, native people of Arctic Canada, Greenland, and the United States, should also be considered. The traditional Inuit diet was not only heavy on meat and fat but nearly devoid of plants. Staples included seal, whale, walrus, moose, caribou, reindeer, ducks, geese, quail, salmon, whitefish, tomcod, pike, char, and crab. Seal oil was used for cooking and as a sauce.[136] "[T]he traditional Eskimo diet had little in the way of plant food, no agricultural or dairy products, and was unusually low in carbohydrates." Modern analyses of Inuit health are confounded by various factors, including astronomically high smoking rates (79 percent). Nonetheless, studies show that Inuits whose diets are the closest to the traditional diet—those consuming large amounts of animal flesh and fats, including saturated fat—are the least likely to succumb to heart disease.

While there is limited written record on the health of early Native Americans, evidence suggests that their health was generally excellent and free of chronic diseases. Early explorers, including 16th-century Spaniard Cabeza de Vaca, wrote admiringly of the exceptional vigor of Native Americans. "The men could run after a deer for an entire day without resting and without apparent fatigue," he said. Dr. Weston Price, a dentist who, in the 1930s, spent considerable time among native communities in various parts of the world, "noted an almost complete absence of tooth decay and dental deformities among native Americans who lived as their ancestors did."[137] Price also interviewed health care practitioners, including a Dr. Romig in Alaska, who reported that "in his thirty-six years of contact with [native] people he had never seen a case of malignant disease among the truly primitive Eskimos and Indians, although it frequently occurs when they become modernized."

Complementing historical accounts are archaeological investigations, particularly examinations of human bones and teeth. Such investigations have found that the major chronic health problem in ancient populations in the New World was iron deficiency anemia.

But the same evidence suggests the anemia was the result of abandoning the hunter-gatherer lifestyle and increasingly relying on agriculture, specifically maize cultivation. "[I]ncreased population density [from the development of agriculture in North America], along with other ecologic and demographic changes associated with intensified farming, had a profound influence on health, with statistically significant increases in infectious diseases and iron deficiency anemia."[138]

Looking outside the Americas, the Masai have long been a subject of fascination to Westerners. A livestock-herding people living in Kenya and Tanzania, the Masai are living contradictions to the prevailing anti-fat, anti-red-meat dogma found in the West. "Their diet," a 2012 *Science Nordic* article notes, "is as full of fats as the diet of people living in the West."[139] Historical accounts by Moritz Merker, a German who lived among the Masai for eight years, beginning in 1895, suggest the traditional Masai diet has long consisted mostly of dairy and red meat. "The favorite foods of the Masai, Merker tells us, were milk, meat, and blood. They would drink the blood of cattle, sheep, or goats in the fresh or clotted state."[140] "Unlike Westerners, however, the Masai do not have many problems related to lifestyle diseases."[141] New research shows that, counter to previous claims, the good health of the Masai is not due to intense physical activity. Instead, the Masai engage in "moderate but constant physical activity." Regardless, the lack of chronic diseases among the Masai strongly suggests that there is no property inherent to saturated fat and red meat that causes heart disease and other chronic diseases.

The diets of the Masai, and early Inuits and other Native Americans, rich in red meat and fats, including plenty of saturated fat, demonstrate that there is nothing contained in those foods that per se raises risk of heart disease or other chronic diseases. "Dr. Price found fourteen groups—from isolated Irish and Swiss, from Eskimos to Africans—in which almost every member of the tribe or village enjoyed superb health. They were free of chronic diseases . . ."[142] In *Guns, Germs, and Steel*, Jared Diamond writes of the paradox that "food production, while increasing the quantity of edible calories per acre, left the food producers less well nourished than the hunter-gatherers whom they succeeded."[143]

Cheap Food Is Cheap

Sugar and flour. We've made them cheap and abundant and now we consume them in great excess. Our collective health is paying the price. In our modern culture of abundance, we no longer regard or treat food generally as something precious. We expect it to be cheap, and it is. Not too long ago we spent a lot more of our income on food. In 1933, we spent 25 percent of our income on food; in 1950 this was up to 31 percent. But in 2011 we spent just 9 percent. Not surprisingly, the amount we spend on food runs parallel with our attitudes toward it: That which we can obtain cheaply, we expect little value from.

For years, America has massively overproduced grain and other commodities. People often assume that factory farms and feedlots are necessary to feed the growing population, but this belies the historical facts. In their treatise on agriculture, *American Farm Policy 1948–1973,* professor of agriculture Willard Cochrane and agricultural research fellow Mary Ryan note: "Except in wartime there have been too many resources producing too much product—excess production capacity—on American farms since 1930; this is the basic problem of the commercial farm sector."

At the same time, we waste more of our food than ever before. Nearly half of all food produced in the United States is never eaten. By weight, food waste is the single largest component of American garbage. Recovery of just 25 percent of wasted food would feed 20 million people, the USDA estimates.

We exacerbate the situation by subsidizing the wrong foods. Sugar, nutritively the least valuable and among the most damaging to our health, is subsidized to the tune of $4 billion per year. Corn (used both as a grain and to make high-fructose corn syrup) has been subsidized at $80 billion over the past 15 years. This is part of the 45 percent of subsidies that are funneled into feed grains, especially soy and corn. An additional 30 percent of U.S. agricultural subsidies go to wheat and rice.

The current industrialized food production system leads us to eat an abundance of unhealthy foods. They are produced in excess and we overeat them. With cheap food we get what we pay for.

CHAPTER 8

BEEF IS GOOD FOOD

HOPEFULLY BY THIS POINT YOU ARE OF THE MIND that beef has been wrongfully accused of being bad for your health. But still you may wonder, why eat it? We've been so deeply trained in recent decades to mistrust both the environmental soundness of cattle ranching and the healthfulness of beef that perhaps you've gotten in the habit of avoiding it or eating it only on rare occasions.

Even so, most people are, at least occasionally, strongly drawn to a hamburger or steak off the grill, a roast beef or pastrami sandwich, or spaghetti with meatballs. Other than brewing coffee, or frying bacon, I find few smells more tantalizing than beef on the grill. Just a few days before I wrote this, a friend of mine who is a longtime vegetarian posted a Facebook confession that she'd given in and eaten a beef hamburger the night before.

In mulling over the question, "Why do people love meat?" Harold McGee explains it like this:

[T]he deepest satisfaction in meat probably comes from instinct and biology. Before we became creatures of culture, nutritional wisdom was built into our sensory system, our taste buds, odor receptors, and brain. Our taste buds in particular are designed to help us recognize and pursue important nutrients: we have receptors for essential salts, for energy-rich sugars, for amino acids, the building blocks of proteins, for energy-bearing molecules called nucleotides. Raw meat triggers all these tastes, because muscle cells are relatively fragile, and because they're biochemically very active . . . Meat is thus mouth-filling in a way that few plant foods are. Its rich aroma when cooked comes from the same biochemical complexity.[1]

Our instincts draw us toward beef while our brains—ringing with alarm bells and warnings that have been crammed in there by modern health advice—may turn us away. Yet we need to let those instincts guide us more in life, and especially when it comes to what whole foods to eat. That deeply inborn impulse we so often ignore or suppress is often leading us in the right direction. Our bodies know that the milk, fat, and flesh of cattle are not only delicious and nutrient-dense, but also uniquely able to provide the human body nourishment in a usable form. "[A]n impartial review of the evidence indicates that red meat is one of the healthiest foods you can eat," writes nutrition expert Chris Kresser.[2]

Theories abound about what we ought to eat and why. For my part, I find the following idea the most credible: We should eat what our bodies evolved to eat. The ancestors of modern humans began consuming the flesh and fat of animals at least 2.6 million years ago, and began eating substantial amounts around 1.5 million years ago.[3] Both because of its rich nourishment and because of the complicated nature of hunting, this shift in eating patterns is closely associated with the increasing size of our brains.

Meanwhile, being omnivores has ranked among our greatest survival advantages. On the opposite end of the spectrum are creatures like koalas, pandas, and monarch butterflies that can live only on a narrow range of foods, thus ensuring extinction if those food sources are wiped out. We humans, by contrast, are highly adaptable. We do well on a vast diversity of plants, animals, and fungi, and our diets have always varied widely depending on climate and geography. The diet for every group of humans around the globe has included a smorgasbord of nature's local offerings.

While I do not believe in any one particular, hard-and-fast dietary dogma (in fact, I find most rigid diets absurd), I consider our bodies to benefit greatly from eating the animals of the field, as our ancestors did for countless generations. Like bison, antelope, moose, alpacas, yak, gazelles, deer, elk, camels, and caribou, cattle are grazing animals uniquely able to eat grass and other cellulosic plants from which humans themselves can get no direct nourishment. Like other grazing animals, cattle miraculously convert those plants into milk and flesh, which are jam-packed full of nutrients. And in contrast with those in plants, the nutrients in bovine milk and meat are in a form that is often uniquely usable by the human body. (I'll talk more about this momentarily.)

I strongly believe as well that a person should always strive to get nourishment from food. Not pills, powders, or liquids concocted in laboratories—good old-fashioned food. Under particular circumstances, from time to time, supplementation may be helpful or even necessary. But to me, any prolonged reliance on manmade supplements rather than food is a sign of an unhealthy diet.

Experts widely agree that food is the best source of human nourishment. The U.S. government's Dietary Guidelines note that "nutrients should come primarily from foods."[4] The Mayo Clinic website points to three primary benefits of foods, compared with nutritional supplements: greater nutrition ("Whole foods are complex, containing a variety of the micronutrients your body needs"); fiber; and protective substances ("whole foods contain other substances important for good health," including phytochemicals and antioxidants).[5] Dr. Frank Hu, Harvard professor of nutrition and epidemiology, urges that looking at food merely as component nutrients is a simplistic and "outdated" approach. He stresses the importance, instead, of eating "real food."[6] Increasingly, eating a diversity of whole, unprocessed foods is regarded as the key to good nutrition. Dr. Yudkin would have been pleased.

Despite the avalanche of books and articles written in recent years about what to eat, nutrition science is in its infancy and much remains unknown. To me, this ignorance is the most compelling reason to aim for a diet rich in nutrients that contains everything a human body needs. We simply don't know enough to know whether dietary supplements can ever adequately replace essential nutrients our forebears always derived from foods. New York University nutrition professor Marion Nestle notes: "Clinical trials rarely show much benefit from taking [nutritional] supplements and . . . sometimes they show harm."[7]

My own diet has long been founded on fresh and fermented foods—well raised, locally sourced, and simply prepared. In recent years, I have increased eggs and dairy products (both from farms where the animals are on pasture) while cutting back substantially on both sugary and starchy carbohydrates. In late 2013, I stopped eating sucrose and all other sweeteners entirely. At the same time, I eat more fruits and vegetables than ever. I include plenty of fats and fatty foods—including nuts, avocados, olive oil, coconut oil, and butter—and I consider them essential to good health.

While I consume ample bovine fat in the form of milk, yogurt, cheese, and butter, I still do not eat meat. This may surprise (and even annoy) some people. How can I encourage others to eat beef when I do not do it myself? By way of explanation I can only say this: I stopped eating meat over two decades ago, at a time when I believed it was the right thing for the environment and my health. Over the past 12 years my views have shifted dramatically. I now view animals as an essential part of an environmentally optimal food system. And I consider the ideal diet to include meat, and definitely beef. But as is the case for everyone, multiple considerations enter into my daily choices about what to eat. And though I recognize that my diet is less than optimal because it does not include meat, to date I simply have not had to urge to eat it. If I ever regain the desire to eat meat, I will.

Evidently, I respect the choice to abstain from beef. But if your rationale is based on the environment or health, I think the reasons are poorly grounded. As I argued in the first half of this book, nothing about raising cattle is inherently damaging to the environment. Ecological injury from cattle tending results from mismanagement. And as I argued in the preceding chapter, fears over negative health consequences from eating beef are proving unfounded. The benefits to human health, on the other hand, are well documented.

A 2013 article in *The Guardian* calls beef "one of the most nutritious foods" available, noting that "beef has appetite-sating high-quality protein, which has all the essential amino acids needed (isoleucine, leucine, lysine, methionine, phenylalanine, threonine, tryptophan, valine and more) to build muscle and bone." Beef, it points out, is also a great source of B vitamins, iron, and zinc.[8] To that list I would add vitamin D, found in very few foods, yet available in the most bio-available form in beef liver.

As I'll discuss more a bit later, the most nutritious and wholesome beef you can eat is from cattle raised entirely on grass and other forages. Yet even standard grocery store beef is crammed full of nourishment. It takes just a 3-ounce piece of grain-finished beef to provide the following part of the U.S. recommended daily allowance for an adult male:

Protein: 23 grams (45% of USRDA)
Minerals: 13% iron; 5% magnesium; 42% zinc; 4% copper; 20% phosphorus; 24% selenium
B vitamins: 5% thiamin (B_1); 10% riboflavin (B_2); 16% niacin (B_3); 14% pyridoxine (B_6); 54% cyano-cobalimin (B_{12}).

Stepping it up, an exceptionally nutrient-dense meal can be made of beef organ meat, for which just 3 ounces includes the following:

Protein: 21 grams (43% USRDA)
Vitamin C: 71% USRDA

Minerals: 186% iron; 4% magnesium; 16% zinc; 39% copper; 26% phosphorus; 7% potassium; 111% selenium

B vitamins: 3% thiamin (B_1); 15% riboflavin (B_2); 24% niacin (B_3); 7% pantothenic acid (B_5); 2% pyridoxine (B_6); 71% cyano-cobalimin (B_{12})[9]

As anyone watching their weight these days knows, either of those pieces of beef will also contain zero carbohydrates, and rank zero on the glycemic index. These latter qualities make beef an exceptionally good choice for dieters and diabetics. Although, for reasons discussed previously, I do not think saturated fat is a health concern, many people will be interested to know that the steak contains just 16 percent of the daily saturated fat allowance, and the organ meat contains just 6 percent. Truly, beef is among the most nutritious foods available.

Nutrients in Beef, Uniquely Bio-Available

Now to dig a little deeper into beef's benefits. Let's start with one of the most obvious: iron. As the nutritional information just presented makes clear, beef is loaded with iron. A small steak has 13 percent of the recommended daily allowance, while organ meats have a whopping 186 percent. Even the steak is higher than lamb, more than double turkey, and far more—three to four times more—than any other mainstream meats. As I'll explain momentarily, the iron found in meat is also uniquely usable by the human body.

"Iron deficiency is the most common and widespread nutritional disorder in the world," according to the World Health Organization. Up to 80 percent of the world's people is estimated to be iron-deficient, with 30 percent of the world's population afflicted by full-blown anemia. Women and children are hardest hit, especially pregnant women. In developing countries, half of pregnant woman and about 40 percent of preschool children are estimated to be severely iron-deficient. At the same time, WHO says lack of iron is "the only nutrient deficiency which is also significantly prevalent in industrialized countries."[10]

In the United States, iron deficiency is widespread among girls and women, especially those of lower income and those who are pregnant. One study showed that more than 80 percent of women 20 to 49 years of age were failing to get recommended amounts of iron from their diets.[11] The National Institutes of Health reports that "anemia among lower income pregnant women has remained the same, at about 30 percent, since the 1980s."[12] More than three million Americans, mostly women and people with chronic diseases, have chronic anemia.[13]

Anemia is a serious condition in which the body has fewer-than-normal red blood cells in the blood, or the red blood cells have insufficient hemoglobin (the protein that carries oxygen from the lungs to the rest of the body). In the short term, iron deficiency causes low energy levels, and hinders memory and other mental functions.[14] Longer term, the medical consequences of anemia can be severe, especially for pregnant women. For all people afflicted, it can impair organ function. For pregnant women, it can impair physical and cognitive fetal development and contributes to 20 percent of all maternal deaths.

Women of child-bearing age need the most iron in their diets (the U.S. recommended daily allowance is 18 milligrams for a woman between 18 and 50), with pregnant women need nearly twice that (around 30 mg).[15] Women are the most at risk for iron deficiency.

The problem is less in finding foods containing iron than it is in the human body's ability to utilize the iron. "Note that iron intake is not equal to iron absorption," advises the Cleveland Clinic.[16] Iron in food comes in two forms: heme (which is bound to hemoglobin) and nonheme (which is not). Meat, especially red meats, contains large portions of heme, whereas plants that contain iron, such as lentils and beans, have only nonheme iron. It's also important to note that nonheme iron is the type added to iron-enriched and iron-fortified foods. Although most dietary iron is nonheme, "heme iron is absorbed better than nonheme iron," according to the National Institutes of Health.[17] The NIH advises that vegans and vegetarians need twice as much dietary iron each day as meat eaters because of the poor absorption of iron from plants. "Absorption of iron into the body is greatest with meat sources of iron," Cleveland Clinic also advises.[18]

More specifically, the NIH notes that "absorption of heme iron from meat proteins is efficient" and is not much influenced by other foods one eats. In contrast, NIH says, only "2 percent to 20 percent of nonheme iron in plant foods such as rice, maize, black beans, soybeans and wheat is absorbed," and its "absorption is significantly influenced by various food components." Meat not only provides a lot of iron, and better-quality iron, but also assists in the absorption of nonheme iron, which is actually, ironically, hampered by some vegetable protein sources, including soybeans.[19] "The reason for this is not well understood," notes Harold McGee about the differing absorption rates.[20] But clearly, a person who regularly eats red meat is far less likely to have an iron-deficiency problem than someone who does not.

The challenge of getting sufficient iron without red meat became real and personal to me in my pregnancies. Both times, blood tests revealed my iron levels to be dangerously low. In the first pregnancy, I was able, with a daily prenatal supplement, and extremely carefully planned eating, to raise my iron levels to just barely within the acceptable range. In the second pregnancy, however, even with what seemed like Herculean efforts on my part, plant-based foods and prenatal vitamins could not raise my body's iron to a safe level. Once I began adding a daily liquid iron supplement, it was just barely enough. It was fortunate I took the issue seriously. Just after my second son was born, I had post-delivery bleeding that necessitated emergency surgery. The hemorrhaging was severe enough that, doctors later informed me, had I not been at a modern medical facility I might well have died. The iron my body had stored in the preceding months was important in my survival and recovery. Had I been eating red meat, however, I probably would never have faced an iron shortage at all. My advice to any non-vegetarian pregnant women is to be sure to regularly include some beef in their meals.

An interesting side note about iron, which seems to follow the general pattern that what we evolved with works best in the human body: Iron in breast milk is far better absorbed by infants than iron in manmade formula. "It is estimated that infants can use greater than 50% of the iron in breast milk as compared to less than 12% of the iron in infant formula," according to NIH.[21] "Reasons for the high bioavailability of iron in breast milk are unknown," states a hematology textbook.[22] (Further proof of how little we understand about human nutrition, and yet another important reason to breast-feed.)

As the nutritional information presented above makes clear, just a 3-ounce serving of steak provides nearly half an adult's zinc needs. Even compared with other red meats, beef is loaded with zinc: it contains about a third more than is found in lamb, and more than twice as much as in veal and pork.

The National Institutes of Health advises that zinc, which is present in cells throughout the body, is essential to good health. It's important in immune system function, protein and DNA synthesis, and—in pregnancy, infancy, and childhood alike—zinc is essential for development, according to NIH. Zinc aids in the healing of wounds and is important for functioning of the senses of taste and smell.[23]

Globally, zinc deficiency is widespread. By some estimates, it is also common in the United States. One study showed that 40 percent of U.S. adolescent girls and adult women were failing to get the recommended daily allowance of zinc, while another showed that 60 percent of men over the age of 70 were failing to get adequate zinc.[24]

As with iron, the zinc found in beef is especially usable by the human body. According to the United Nation's Food and Agriculture Organization, studies show that about twice as much zinc is absorbed from diets containing meat compared with those without meat. This is because plant foods with high levels of zinc (including legumes, whole grains, seeds, and nuts) also are high in phytate, a substance that inhibits mineral absorption. Just as with iron, the availability of zinc from plant foods can be improved by inclusion of animal protein sources, FAO notes.[25]

Food for Our Bones, Muscles, and Brains

Protein is another good reason to eat beef. As noted above, a small portion of beef provides nearly half the recommended daily allowance of protein. In the developing world, protein deficiencies are widespread. Generally speaking, most Americans are not protein-deficient. Still, in all countries various situations warrant a diet richer in protein and fat and lower in carbohydrates. As discussed previously, lowering carbohydrates helps control diabetes and assists in weight loss. Additionally, as people age, meat protein seems to become even more important to good health. As with iron, the protein contained in beef is of a particularly high quality, which appears to become more important as we age. A study conducted by researchers in Ohasama, Japan, and published in the March 2014 *Journal of the American Geriatrics Society* found that "after adjustment for putative confounding factors, men in the highest quartile of animal protein intake had significantly lower risk of higher-level functional decline than those in the lowest quartile." The study found no benefit to either gender from plant protein. The study authors conclude: "Higher protein, particularly animal protein, was associated with lower risk of decline in higher-level functional capacity in older men."[26]

Likewise, because we tend to lose muscle mass as we age, and our bones become weaker, high-quality protein, like that provided by beef, becomes more important as we get older. A 2008 study in the *American Journal of Clinical Nutrition* notes that "[m]aintenance of adequate bone strength and density with aging is highly dependent on the maintenance of adequate muscle mass and function, which is in turn dependent on adequate intake of high-quality protein."[27] The researchers conclude that despite some belief to the contrary, "higher protein diets are actually associated with greater bone mass and fewer fractures when calcium intake is adequate."

The study's authors urged the retraining of medical professionals about the importance of protein in diets of older patients.

Similar results were obtained by researchers at Hebrew Senior Life's Institute for Aging Research (IFAR), which is affiliated with the Harvard Medical School. In 2000, it published a study in the *Journal of Bone and Mineral Research* concluding that protein intake generally, as well as animal protein in particular, promotes bone health. The researchers found that lower protein consumption was significantly related to bone loss, "suggesting that protein intake is important in maintaining bone or minimizing bone loss in elderly people." An IFAR article on the study notes that the research conformed with two earlier randomized controlled trials showing that increased protein intake dramatically improved outcomes after hip fractures.[28]

For people of all ages, beef provides plenty of nutrients valuable to optimal brain function. As a 2008 article about "brain food" in *Nature* makes clear, beef contains the following nutrients involved in brain function: iron, copper, selenium, zinc, B_6 and B_{12} vitamins, choline, vitamin D, and vitamin C.[29] Infants and children need cholesterol, like that contained in bovine fat, for proper brain development.[30] Cattle raised entirely on grass also have vitamin E, omega-3 fatty acids, and beta-carotene in their flesh, fat, and milk.[31]

Best Food Source for Vitamin D

In recent years, an enormous amount of attention has been paid to vitamin D deficiencies. "In the United States and Europe, it is estimated that more than two-thirds of the population is deficient in vitamin D," *The New York Times* reported in 2014.[32] Increasingly, vitamin D shortfalls are being connected to everything from diabetes, stroke, hypertension, and autoimmune disorders to heart disease.[33]

Part of the challenge is that vitamin D is found in few foods. The most usable form of the vitamin comes from time spent in the sun. For most of the year, in most latitudes where humans reside, you can get adequate vitamin D merely by spending brief amounts of time outside every few days. The epidemic nature of today's vitamin D shortages is yet another side effect of modern lifestyles spent indoors in front of computers and TV screens along with the overly zealous application of sunscreens intended to decrease damage to the skin from UV exposure.

For those who cannot get enough sunshine, some of the best foods available to bolster vitamin D levels are whole milk (especially from cows kept on grass) and beef liver, which can provide 19 IU per 100 grams. This is only a small portion of what you need, but it could be a vital part of a strategy to improve your vitamin D levels. In fact, red meat "contains a vitamin D metabolite called 25-hydroxycholecalciferol, which is assimilated much more quickly and easily than other dietary forms of vitamin D," according to Chris Kresser.[34] There is even evidence that meat (although not milk) provides protection against rickets, a disease caused by severe vitamin D deficiency, suggesting that the vitamin D in meat is "uniquely absorbable and useful to the human body."[35]

B as in Beef

As mentioned already, beef is also an excellent source of B vitamins. If you don't understand what's meant by "B vitamins," you are not alone. The answer is complicated and has changed considerably in recent years. Once regarded as a single vitamin, scientists now consider the B's a complex of 10 or 11 (depending on whether you include choline) chemically distinct vitamins often found in the same foods and frequently functioning together as a group.[36] The B vitamins are now believed to have considerable importance, affecting the functioning of "every organ system and all aspects of our health."[37] In particular, they are thought to be important in methylation, the process by which protein and DNA are produced and sustained in the body; important in heart health; and essential in creating neurotransmitters responsible for brain function and mood.[38]

B vitamin deficiencies were once considered rare. Yet a school of thought is lately emerging that B vitamin deficiencies may be widespread—affecting as much as 60 percent of the U.S. population.[39]

Beef and beef organs are good sources of nearly all the B vitamins, including B_1, B_2, and B_5, as well as PABA, inositol, and choline. The National Academy of Sciences has recommended a daily intake of choline of 550 mg for men and 425 mg for women. Just 3 ounces of ground beef contains 67.4 mg of choline (12 and 16 percent, respectively, of recommended amounts).[40] B_6, a deficiency of which may contribute to serious health problems, including cardiovascular disease and cognitive decline, is found in both beef and beef organ meats.[41]

Vitamin B_{12}, which is available only in animal-based foods, is of particular concern. In certain populations, especially those that eschew all animal products, deficiency is widespread. B_{12} deficiency affects between 1.5 and 15 percent of the American public. Those most at risk are people over age 50, people with pernicious anemia, people with gastrointestinal disorders (including Crohn's disease and celiac disease)—all of whom may have difficulty absorbing B_{12}—along with vegetarians and especially vegans, whose diet is likely to contain inadequate amounts. A 2010 Oxford University study showed that over half of vegans and 7 percent of vegetarians were B_{12}-deficient.[42] "When pregnant women and women who breast-feed their babies are strict vegetarians or vegans, their babies might also not get enough vitamin B_{12}," NIH advises.

Infants who don't get enough vitamin B_{12} may experience developmental problems and megaloblastic anemia (a disorder characterized by large but immature and incompletely developed red blood cells that do not function like healthy cells). In adults, B_{12} deficiency can cause tiredness, weakness, depression, dementia, and megaloblastic anemia. Up to 30 percent of patients hospitalized for depression are B_{12}-deficient.[43] Long term, the deficiency can damage the nervous system. Among otherwise healthy people who regularly eat meat, B_{12} deficiencies are uncommon. All beef contains B_{12}, and beef liver is one of the best sources of the nutrient, according to NIH.[44]

By now you have undoubtedly noticed a pattern: Beef organs stand out as foods especially packed with nutrients in a highly usable form. In short, they are one of the best sources of some of the most vital nutrients in the human diet. Integrative health practitioner Chris Kresser, who calls organ meats "nature's most potent super-foods," points out that in some traditional cultures, *only* the organ meats were consumed while the lean muscle meats we mostly eat in the United States today were discarded or given to the dogs.[45] As a general rule, Kresser says, "organ meats are between 10 and 100 times higher in nutrients than corresponding muscle meats." He also notes that while people often worry that environmental toxins might build up in cattle livers, in fact they do not. He advises his patients to "eat meat and organ meats from animals that have been raised on fresh pasture without hormones, antibiotics or commercial feed. Pasture-raised animal products are much higher in nutrients . . ."

Comparing Grass-Fed with Grain-Fed

While this is certainly true for pasture-raised eggs, when it comes to beef and dairy products, Kresser's comment is a bit of an over-statement. (I'll get more specific on that point in a moment.) The reasons to seek out dairy and meat from cattle raised on grass are compelling, nonetheless. As I will discuss later, the majority of cattle raised for beef in this country are implanted with growth hormones (usually both on the ranch and again at the feedlot), and most are also fed antibiotics and other pharmaceuticals as part of their feedlot ration. In contrast, I have never heard of a farm or ranch raising cattle for truly grass-fed beef that engages in either practice. To me, when combined with the environmental problems associated with large feedlots discussed in the first half of this book, the reasons for seeking foods from cattle raised wholly on grass are compelling.

I strongly favor grass-based farming and ranching. It's more ecological. It creates a more humane, healthy living environment for animals. It's a better work environment for the humans tending the animals. It generates healthier, safer food.

Life on the pasture provides manifold benefits even for omnivorous animals like turkeys, chickens, and pigs.[46] I have long argued against continual indoor confinement of farm animals: All creatures need daily exercise, fresh air, and sunshine to be healthy and have a decent life. Not surprisingly, animals must have a decent life if they are to provide flavorful food.

"Rapid, confined growth favors the production of white muscle fibers, so modern meats are relatively pale," writes famed food scientist Harold McGee.[47] "They're also tender, because the animal gets little exercise, because rapid growth means that their connective tissue collagen is continuously taken apart and rebuilt . . ." On the other hand, "Full-flavored meat comes from animals that have led a full life," McGee notes. "[W]ell-exercised muscle with a high proportion of red fibers (chicken leg, beef) makes more flavorful meat than less exercised predominantly white-fibered muscle (chicken breast, veal) . . . This connection between exercise and flavor has been known for a very long time."[48] McGee also notes the connection between fat and flavor, and that the modern tendency to slaughter animals younger and lighter has diminished eating quality. "Thanks to the industrial drive toward greater efficiency," he notes, "and consumer worries about animal fats, meat has been getting younger and leaner, and therefore more prone to end up dry and flavorless."[49]

Feeding antibiotics or other drugs to cattle, as well as using growth hormones, causes environmental contamination and raises the specter of food tainted with drugs, hormones, and antibiotic-resistant bacteria. There is evidence that keeping cattle on forages (rather than concentrated feeds like grain) up to the point of slaughter can make beef safer, by reducing the occurrence of the dangerous *E. coli* strain 0157.H7.[50]

Nonetheless, I want to take a moment to correct a couple of modern myths surrounding cattle feeding. Contrary to what's often suggested in the popular press these days, there's nothing inherently wrong with feeding cattle a certain amount of grain. As discussed earlier, modern cattle breeds—for both meat and milk—are descended from a long line of domesticated cattle fed grain at least some of the time. Cattle that are sufficiently mature and are given grain in appropriate amounts are well equipped to tolerate grains in their diet.

Keeping all of that in mind, I consider grass the optimal diet and environment for cattle. It provides animals by far the most natural life and creates the safest, highest-eating-quality, and most nutritious food. The mainstream beef industry is fond of funding health and environmental studies purporting to show that there's no difference between typical feedlot beef and grass-fed beef. However, the scientific consensus says otherwise. It is true that numerically the differences in nutrient content are not large. Yet there is a consistent, across-the-board nutritional advantage to beef and milk from cattle raised entirely on grass and other forages.

To date, the most comprehensive review of studies comparing nutrition of foods from grass-fed versus grain-fed cattle was done in 2006 by Union of Concerned Scientists senior scientist Kate Clancy, who holds a doctorate in nutrition science. Regrettably, the report, titled *Greener Pastures*, is largely focused on fats, which, clearly, I do not consider a health concern. Nonetheless, the report is credible and important. Overall, it concludes: "Meat from pasture-raised cattle contains less total fat than meat from conventionally raised animals, and both meat and milk from pasture-raised animals contain higher levels of certain fats that appear to provide health benefits." The nutritional differences, the report notes, result from "the chemical differences between forage and grains, and the complex ways in which ruminant animals such as cattle process these feeds."[51]

More specifically, the report found that all the grass-fed steak in the studies reviewed could be labeled "lean" or "extra lean"; some could also be labeled "low-fat" (no more than 3 grams of total fat per serving).[52] Steak and ground beef from grass-fed cattle are almost always lower in total fat; steak from grass-fed cattle tends to have higher levels of the omega-3 fatty acid alpha-linolenic acid (ALA); and ground beef from grass-fed cattle usually has higher levels of conjugated linoleic acid (CLA). Meanwhile, milk from pasture-raised cattle tends to have higher levels of ALA and has consistently higher levels of CLA. ALA is believed to reduce heart disease risk, while CLA may have positive effects on heart disease, cancer, and the immune system. "CLA has been associated in animal and laboratory studies with an impressive array of health benefits . . . [relating to] atherosclerosis, diabetes, immune function, and body composition."[53]

Greener Pastures also found strong evidence of higher vitamin content in foods from grass-fed cattle. "[A] substantial number of studies that measured alpha-tocopherol showed that meat and milk from pasture-raised animals had levels of vitamin E significantly higher than meat and milk from animals fed grain," the report concludes.[54] It found the same pattern for beta-carotene, the plant precursor of vitamin A. "Forages contain substantial amounts of beta-carotene and grains contain very little," the report explains. "[B]ecause beta-carotene is easily oxidized, stored forages have a significantly lower concentration of the substance."[55] These factors make the forage-based diet healthier for cattle and result in higher levels of both vitamin E and beta-carotene in foods. However, both nutrients are only present in small amounts in beef and milk, even from animals raised on grass.

In the years since the *Greener Pastures* report was produced, a wealth of new research has considered the possible benefits of raising cattle entirely on grass. For the most part, the research has reached conclusions remarkably in keeping with that report. Research continues to show that grain feeding has a negative effect on the presence of omega-3s in cattle fat. This is because 60 percent of the fatty acids found in grass are omega-3s.[56] When cattle are raised entirely on grass, their fat is about 3 percent omega-3 fat.

Research done by Clemson University and the Department of Agriculture in 2009 found that compared with grain-finished beef, grass-fed beef is higher in CLA and total omega-3 fatty acids; it also has a far better ratio of omega-6 to omega-3 fatty acids (1.65 for grass-fed, compared with 4.84 for grain-fed beef). The researchers also found that grass-fed beef has higher levels of each of the following nutrients: beta-carotene, vitamin E (alpha-tocopherol), calcium, magnesium, potassium, and B vitamins thiamin and riboflavin.[57]

A 2011 study in the British *Journal of Nutrition* showed that even moderate consumption of grass-fed meats for just four weeks benefited the omega-6:omega-3 ratio of the human subjects.[58] However, the omega-3 level steadily drops the longer the animal is in a feedlot. By day 196, the level reaches zero.[59]

A 2010 study done collaboratively by California State University and the University of California and published in *Nutrition Journal* looks closely at the ratio of omega-6 to omega-3 fats.[60] It notes that while a healthy diet should consist of roughly one to four times more omega-6 fatty acids than omega-3 fatty acids, "[t]he typical American diet tends to contain 11 to 30 times more omega-6 fatty acids than omega-3s." The report presents evidence of significant differences in omega-6:omega-3 ratios between grass-fed and grain-fed beef. Considering all studies reviewed, it found an overall average ratio of 1.53 for grass-fed beef and 7.65 for grain-fed. Thus, the grass-fed beef has a much more favorable omega-6:omega-3 ratio.

The California study also considered glutathione (GT), which it calls "a relatively new protein identified in foods" that "functions as an antioxidant." GT works within cells, quenching free radicals and protecting cells from oxidized lipids or proteins and preventing damage to DNA. Fresh vegetables and beef are both high in GT, the report notes, and "[b]ecause GT compounds are elevated in lush green forages, grass-fed beef is particularly high in GT as compared to grain-fed . . . Grass-only diets improve the oxidative enzyme concentration in beef, protecting the muscle lipids against oxidation as well as providing the beef consumer with an additional source of antioxidant compounds."[61]

Really Tasty Food

In the final analysis, whenever we consider what to eat we have to come back to the critically important issue of taste. At least that's true for both my husband, Bill, and me. No matter how healthy grass-fed cheese or beef is, who wants to eat it if it doesn't taste great? This is an underlying principle for all the foods we bring into our home and feed our family and ourselves. It's much of the reason we decided to stop buying out-of-season fruits and tomatoes. When Bill first considered returning to raising cattle entirely on grass (as he'd done decades earlier), he took small, tentative steps toward the venture. He wasn't going to do it at all if he could not find a way to consistently make beef that, as he says, "eats great, every time."

He had feasted on mouthwateringly delicious grass-fed beef while traveling in Argentina, Australia, South Africa, and New Zealand. It was rich, flavorful, and with just the perfect chew. But much of the grass-fed beef he'd sampled in the United States simply did not stack up. It tended to be dry, tangy in flavor, and tough. One occasion at a white-tablecloth restaurant in the Midwest particularly stands out in my memory. The chef had personally come to the table to proudly serve Bill grass-fed beef from the Scotch Highlander cattle of a local farmer. Once we were alone Bill turned to me and whispered, "This is inedible." He choked it down, of course, but merely to prevent any hurt feelings. In recent years, though, he's had many more favorable experiences with grass-fed beef in several regions of the country.

The experiences of *New York Times* food writer Marion Burros advanced along a similar trajectory. "When I wrote about grass-fed beef in 2002 there were about 50 producers, and most of what they raised was not very good," she has written. By 2006, however, she found the grass-fed beef sector to have "taken giant steps." Burros could locate about 1,000 farmers and ranchers raising grass-fed beef, and, from an eating-quality vantage point, she felt "more of them are learning to get it right." Her verdict: "My tasting showed that with 100 percent grass-fed beef you can have it all: sustainable, more nutritious beef with clean, juicy, beefy flavor. (Because the beef has less fat, though, it must be cooked at lower temperatures and for less time.)"[62]

The idea that grass-fed beef, especially the way it is generally raised in the United States today, must be approached differently in the kitchen probably has some validity. Food writer Lynne Curry's book *Pure Beef* is based on this premise. Curry takes issue with the standard cooking advice for grass-fed beef (which she sums up as follows: "Don't cook it over high heat!" "Don't cook it past medium rare!" and

"Don't salt the meat before cooking, or it will be dry!"). Nonetheless, she concludes that beef raised entirely on grass requires some special care. Her advice is this: Season it well; don't hesitate to brown it whenever you'd like; use an appropriate cooking method; and cook it shy of the serving temperature you want.[63] Curry does acknowledge, however, that grass-fed beef varies, and that some ranchers have divined "the magic formula for producing pasture-raised beef that is out of this world." Lynne does not mention anyone by name, but I like to think she (who has been to our ranch and eaten our beef) includes the beef from our ranch in that elite group.

Here, for which I apologize in advance, I must brag a little. My husband, Bill, is as hardworking a person as you will ever meet. His passion for his work runs in his marrow. Nothing is ever good enough unless it's perfect. Speaking from the perspective of his spouse, I can tell you this has both advantages and disadvantages. But the people who always benefit from Bill's relentless passion for excellence are his customers.

After having tasted grass-fed beef from around the United States and in various distant countries, Bill had a realization: When it was bad, grass-fed beef was very bad indeed; but when it was good, it was the best of all beef. In the late 1990s, he began giving serious consideration to raising totally grass-fed cattle himself. When I first met him in 2000, Bill was entertaining the notion of developing a line of totally grass-fed beef for Niman Ranch, the company of which he was founder and CEO. He was being egged on somewhat by things he was reading and by conversations he was having with environmental advocates, writers, and chefs. Foremost among them was Alice Waters, who is not only a world-famous restaurateur but a longtime customer of Bill's who had become convinced that all beef should be raised entirely on grass. Yet like Bill, Waters was finding the eating quality highly variable, and based on her years working with Bill she believed he could produce excellent grass-fed beef if he set his mind to it.

As Bill and I became acquainted, I encouraged him in this pursuit, as well. I had read and seen enough to know that raising cattle entirely on grass had myriad environmental advantages.

Bill's only hesitation (periodically reinforced) was his concern about eating quality. He was determined to get involved with raising grass-fed beef only if he could *consistently* generate beef that tasted at least as good as beef from cattle finished on grain. This was the genesis of a several-years-long quest. Before getting started, he'd queried longtime ranchers about what they thought it would take to make great beef entirely on grass. And he spoke with chefs about their expectations and experiences on the meat side.

Over the next seven years, Bill raised, every year, a handful of cattle entirely on our Bolinas ranch and exclusively on grass. Each year, he slaughtered them progressively, at different moments, according to the condition of the grasses and the body conditions of the animals. He took the cattle in small groups (just two or three at a time) when he considered each individual animal to be at its peak. Taking careful note of the animal's condition, he observed how it corresponded to differences in the fat and meat. As always, he personally transported each animal to the slaughterhouse and personally oversaw the entire kill and butchering.

Together with his colleagues at Niman Ranch, he would then methodically taste a variety of cuts from each of the carcasses. Additionally, each year, he organized tastings by chefs whose palates he particularly trusted.

His interviews and experiences led to the development of a method. To some degree, we consider this method unique. At the same time, we recognize that what we are doing is really just following age-old practices of good animal husbandry and long-standing knowledge about when animals should be slaughtered. We are happy to share what we have learned with anyone who asks.

Making Great Grass-Fed Beef

Here are the basics of our method. Starting with the right breeds and lineages is essential. We have a herd of mother cows and bulls, all of British breeds (called native cattle in the trade). Every few years we have some Hereford bulls, but our mother cows (and most of our bulls) are predominantly Aberdeen-Angus. After having sampled meat from literally thousands of carcasses over several decades while at Niman Ranch, Bill is firmly convinced that British cattle breeds consistently provide meat of the highest eating quality. As he says, "The British have always been a beef-loving culture. They focused on creating breeds that produced great-eating beef on grass." On top of that, Angus cattle are easy to work with: They are even-tempered, make excellent mothers, have few health problems and no horns.

Our cattle do their coupling in the winter, from mid-December to mid-February. They mate naturally and always seem to enjoy themselves immensely in the process. (If you think I am kidding, you should watch a bull nuzzling and licking a cow's neck in the breeding season.) In a typical year, over 95 percent of our cows become pregnant. Our goal is always to include in our herd only mother cows that need no assistance in birthing and only bulls that sire calves sized appropriately to make that possible.

A calf's gestation period is nine and a half months, so the calves are born the following fall—from September to December. We deliberately time the calving this way to be in harmony with our grasses. Mother cows need more nutrition as their young ones grow, and around here grasses are strong starting in late winter and into the spring. (Conversely, we have found that when there is too much feed available for the mother cows at birthing time, the calves are more likely to get scours and the cows are more likely to have udder problems.)

Calves live with their mothers and the rest of the herd of mother cows for seven to nine months. (It truly takes a cow village to raise a calf. Mother cows collectively watch over the young, especially when it comes to protecting them from predators.) Calves are then separated from their mothers using a "fence line weaning" technique, which keeps the pairs in the same location separated only by a wooden fence, minimizing stress for both cow and calf. From that point forward, the weanlings live together as a group and are always kept on the best pastures due to the high nutritional demands of their growing bodies.

These offspring (other than those heifers we decide to keep, who eventually join the cow herd) will continue to live on pastures and rangelands as yearlings and two-year-olds. More precisely, our cattle will go to slaughter averaging between 24 and 30 months of age and, for steers, weighing between 1,250 and 1,350 pounds; for heifers, between 1,150 and 1,250 pounds.

Anyone who raises cattle will recognize those weights as indicating fully grown, fat cattle. We refer to them as "grass-fattened," and they truly are. Bill often compares the way cattle bodies deliberately fatten themselves, when the opportunity presents itself in abundant feed, to get through the lean times to the way a bear gorges itself on salmon to get through a winter in hibernation.

We consider both maturity and fat vital to creating the best beef. Just as the oft-repeated farming saying goes, "Old hens make the best soup," in all species age adds flavor. "According to a standard French [culinary] handbook, *Technologie Culinaire* (1995), the meat of [a beef] animal less than two years old is 'completely insipid' while meat 'at the summit of quality' comes from a steer three or four years of age," Harold McGee points out.[64] (Although I should note that, with a similar idea in mind, for the first two years, Bill slaughtered only animals that were at least three years old, and even some at four years. He found no improvement in the flavor compared with two-year-olds.) As you can discern from the earlier discussion, we embrace the notion that beef fat—especially from truly grass-raised cattle—is a healthful and delicious part of a human diet.

Every geography has a season for beef, just as it has a season for peaches, strawberries, and tomatoes. In California, with our Mediterranean climate, that time is just after the grasses begin to dry out—usually in May to June. In other parts of the United States, it will be shortly after the first hard frost—typically in October. As I wrote about in some detail in *Righteous Porkchop*, feedlots, where cattle are fed corn and other grains, were invented largely for the purpose of smoothing out the swells and dips in supply—eliminating beef's seasonality. Chefs and other eaters of grass-fed beef should keep this principle in mind. Anyone who wants to eat great tasting grass-fed beef (and beef at its nutritional peak) must seek beef from cattle slaughtered in the right season. You can eat beef fresh in that season and frozen (or otherwise preserved) for the rest of the year.

To state it briefly, then, we believe consistently great grass-fed beef requires the following elements: British-breed cattle, kept always on grass, raised to maturity (minimum 24 months), slaughtered just after the grass season has peaked and the cattle are in prime condition (read: fat).

Our cattle, both those raised for meat and the breeding animals, spend every day of their lives on pasture or natural rangelands. The sole exceptions to this are when we have an individual sick animal (who may be kept in a "sick pen" in the corrals until recovered), and during the weaning (when, as I just mentioned, the calves will be kept in the corrals for several days so that they can share a wooden fence line with their mothers, who stay on pasture). Or, in a really good grass year, the cows might be kept in the corrals and the calves on the other side of the fence grazing highly nutritious late-spring feed.

As for the feeding, the regimen is straightforward. We never feed any grain, nor by-products, nor drugs of any kind. (And it goes without saying, we never use any growth hormones.) From birth until weaning, calves nurse on their mothers and graze on pasture. Other than their mother's milk, the only thing our cattle ever eat is grass and other naturally occurring vegetation (like clover and vetch), along with a small amount of alfalfa hay.

The hay we use accounts for less than 1 percent of their annual diet. We'd rather not feed hay at all because it needs to be grown elsewhere, and transported here, which adds to the resources expended, and adds considerably to our input costs. However, the protein (we buy only hay deemed high-quality but that didn't test high enough for dairies, at around 20 percent protein) is a helpful supplement to the mother cows' diets in the dry season, when the nutritional content of our forages is at its lowest. We could operate without hay. But it would mean less vigorous animals, perhaps more susceptibility to disease, and most likely reduced fertility rates. This reduction in herd health and fertility also translates to a greater environmental impact.[65] Regarding alfalfa feeding, the Pimentels note: "In contrast to most other crops, alfalfa needs little or no nitrogen fertilizer; like legumes, it is associated with nitro-gen-fixing bacteria. Because nitrogen fertilizer is an energy-costly input, this savings helps keep alfalfa production relatively energy efficient."[66] We believe that, on balance, this limited feeding of hay, when needed, is the right thing to do.

All of the cattle that go into the beef we sell under own name (BN Ranch) come from our own herd or from the herd of a few other ranches we know well and work closely with, and that follow the same feeding and husbandry practices. We never buy any cattle from sale yards or any other source other than ranchers we know personally. This approach assures us and our customers that the origins and history of all of our beef is fully known and transparent.

A lot of chefs have told us that ours is the best beef they've ever tasted. Not the best grass-fed beef, the best beef period. Internationally renowned chef Dan Barber said that eating our beef was "a transformative experience." So although we continue to look for ways to improve, I think we've found a successful for-mula. We are proud of the way we raise our animals and the beef we produce. We know it is exceptionally nutritious, healthful, and safe, and, equally important, delicious. Why eat anything less?

Beef Lasts

Finally, I'd like to mention one additional aspect, a point often overlooked but important, about why beef is a uniquely valuable food for humans: It's much less perishable than other foods. Until the advent of refrigeration (in 1834), people around the world relied entirely on various techniques like drying, smoking, salting, and pickling to preserve meats and other foods. Some methods enabled meat to be preserved for years.[67] Beef has always been among those foods that stay good the longest.

Somewhat ironically, the very thing for which beef has been most often maligned of late—saturated fats—is much of the reason for its good keeping quality. Harold McGee explains: "Unsaturated fats are most susceptible to rancidity, which means that fish, poultry and game birds go bad most quickly. Beef has the most saturated fat and stable of all meat fats, and keeps the longest."[68]

McGee also explains that "the intact muscles of healthy livestock are generally free of microbes" and that "the bacteria and molds that spoil meat are introduced during processing, usually from the animal's hide or the packaging-plant machinery." In contrast with beef, he notes, "poultry and fish are especially prone to spoilage because they are sold with their skin intact, and many bacteria persist despite washing."

These days, of course, freezing is the primary method of beef preservation, at least in the industrialized world. Butcher and writer Adam Danforth points out in his book *Butchering Beef* that beef is unique in both the quantity of meat provided and the length of time the meat stays good. He points out that the average yield from a whole carcass of beef is more than 400 pounds of boneless meat, making a single animal the source of a year's worth of beef for two average-sized families. "Furthermore, frozen beef stays palatable for a longer period of time than almost any other kind of meat, allowing you to extend your harvest well beyond one year."[69]

One should choose one's beef well and wisely. Then dig in and enjoy.

CRITIQUE

AND FINAL ANALYSIS

WHAT'S THE MATTER WITH BEEF?

GIVEN THIS BOOK'S TITLE, it may surprise some people that I find quite a lot wrong with the ways most cattle in the United States today are raised and turned into meat. Several of these concerns were detailed in the pages in my previous book, *Righteous Porkchop*, and I will not restate all of them here. Additionally, a veritable flood of books and articles has been devoted to describing the "dark sides" of cattle ranching and beef. As I've already made clear, I find a lot of what's been written to range from the poorly informed to the flat-out absurd. But some of the criticisms have validity, and I want to briefly take note of those now.

Stated succinctly, current problems fall into the following categories: the way cattle are managed on the land; substances they are fed; hormones and other drugs used to stimulate growth;

polluting practices; wasted resources; long-distance transport of live animals; and slaughter practices. Said another way, these are problems of land management, wasted resources, pollution, animal welfare, and food safety.

Before becoming directly involved with cattle husbandry, working as an environmental lawyer, I had some distance from these concerns, much as I would in dealing with the pollution of any sector. While, admittedly, I viewed them more objectively, I also had a much shallower depth of understanding. Now, as a member of the community of people who raise cattle and sell beef, I understand the issues far better and, at the same time, feel more urgency about getting them addressed. Each day that problematic practices continue it gives a black eye to everyone who raises cattle or is involved in the production, sale, or preparation of beef.

Every self-help program under the sun says that the first step to correcting your shortcomings is self-examination and acknowledgment. I regularly review several meat industry periodicals and, in nearly every edition, I am struck by the current mode of dealing with consumer concerns, which is never about acknowledgment and always about denial. For every issue that surfaces, it seems that rather than fairly assessing it then rallying the industry to fix it, again and again the meat industry falls back into a circle-the-wagons defensive posture. This has been true for beta-agonists, hormones, antibiotics, and many more concerns. Across the board, there's tone deafness. The industry responds to concerns by saying critics simply "do not understand agriculture," or "don't care about feeding the world," or "need to be educated," and it demonstrates a near-total unwillingness to change or adapt.

Earlier, I cited a series of figures showing that for the past three decades the amount of beef Americans have been eating has been tumbling, a trend that continues to this day. In my view, this industry tone deafness is partially to blame. Survey after survey shows that Americans are becoming less accepting of chemical additives in their food (that's why, for decades, the organic sector has been the fastest-growing segment of the food industry). Yet mainstream cattle nearly across the board continue to be raised with injected hormones, and with beta-agonists and antibiotics added to their feed. At the same time, some of the meat industry's own research shows that people who are reducing their consumption are doing so based on quality concerns. A recent survey in *Meat & Poultry* showed that people who are cutting back on red meat were also seeking red meat of a "higher quality" (whatever that means): "Sixteen percent of Americans who say they are consuming less red meat are now consuming higher-quality red meat."[1] This should be telling the industry that it if wants to remain profitable, it needs to address these concerns. Yet the main initiatives we see from the beef industry continue to be about producing more meat for less cost, which surely *lowers* quality rather than raising it.

A specific example illustrative of the way the cattle and beef industry responds to problems is the recent experience with beta-agonist drugs, including Zilmax. Manufactured by Merck, Zilmax is in of a class of drugs, beta-agonists, that function like steroids. They cause animals to bulk up quickly by converting feed to muscle production rather than fat. First introduced to the market in 2007, the drug was quickly widely adopted by feedlots, and by 2012 it was estimated that 70 to 80 percent of the U.S. cattle herd was being fed Zilmax (or a comparable drug called Optaflexx), generating $159 million in sales for Merck that year. Feedlots favor the drug because the animal gains more meat-producing muscle tissue on less feed.

However, reports of cattle lameness soon surfaced. Tyson Foods began to refuse cattle that had been fed Zilmax after it found that some cattle fed the drug were arriving at its slaughterhouses having difficulty walking. A Reuters investigation found that some lame cattle fed Zilmax were missing hooves. Citing animal welfare concerns, and rejection by Asian markets, Cargill also announced that it would no longer purchase animals fed Zilmax.[2] Astonishingly, none of this prompted the FDA to withdraw its approval of the drug (or do anything at all). But rejection by the big meat players forced Merck, in September 2013, to voluntarily suspended sales of Zilmax in the United States and Canada. Note, though, that Merck has indicated that it does not believe there is any real problem with the drug and intends to compile more "research" to "prove" that the drug is actually safe, so we have certainly not seen the last of it.

This was followed in 2014 by the release of research by a veterinary epidemiologist at Texas Tech University finding that the incidence of death among cattle administered beta-agonists was 75 to 90 percent greater than cattle not administered the drugs.[3] World-famous animal welfare expert Temple Grandin has written and spoken publicly numerous times about the use of beta-agonists in cattle feed and says she believes cattle fed the drugs are "suffering." In one report Grandin wrote: "These observations indicate that there are severe welfare problems in some animals fed beta-agonists."[4] In an interview with National Public Radio, she stated she believed that as many as one of every five cattle fed beta-agonists develop foot problems. Grandin seemed to implore the beef industry to fix the situation when she said: "I've worked all my career to improve how animals are handled, and these animals are just suffering. It has to stop!"[5] China, Russia, and the European Union have already banned the use of beta-agonists in animal feed.

What's been most telling to me is that, despite this flood of negative information, and even trade repercussions, beta-agonists are still used nearly universally at feedlots. Shortly after Merck's withdrawal of Zilmax, *BEEF Magazine* ran a cover story about an agricultural college researcher whose work purports to show that, despite what you may have heard to the contrary, beta-agonists are perfectly safe. Sigh. Tone deafness.

Here's another example: the continual feeding of antibiotics. While the practice is significantly more prevalent at the confinement operations of swine and poultry operations, cattle feedlots feed a lot of antibiotics. They do this both to stimulate growth and to stave off diseases. The Union of Concerned Scientists has estimated that cattle feedlots add about 3.7 million pounds of antibiotics to feed every year, with up to 55 percent of cattle being given subtherapeutic antibiotic doses.[6] In recent decades, evidence has piled up about the dangers of antibiotics overuse with livestock. Due to public health risks, the U.S. Centers for Disease Control and Prevention, World Health Organization, American Medical Association, and American Public Health Association have all taken positions against the practice of continually feeding farm animals antibiotics. Before his death, Senator Edward Kennedy became one of the lead sponsors of the U.S. Senate's version of the legislation. Nonetheless, year after year, the meat industry (including the beef industry) lobbies hard to defeat it; to this day, no law on the subject has passed in the United States. The meat industry continues to claim, too, that where the feeding of antibiotics has been banned, overall use of antibiotics has risen due to increased therapeutic use. This claim has been repeatedly disproven by credible sources, including the Pew Commission on Industrial Farm Animal Production. The facts are very clear: Banning subtherapeutic antibiotics feeding significantly lowers overall usage after the initial year. Despite more than a decade of effort by public health groups and environmental groups, it remains entirely legal in the United States to add low doses of antibiotics to the daily feed and water of cattle, other livestock, and poultry.

Contrast these examples with the way the feeding of antibiotics was handled in Sweden. In the early 1980s, Swedish consumers began to express concern about the overuse of antibiotics in livestock farming. Scientific evidence was starting to emerge that feeding antibiotics to livestock was contributing to the rise of antibiotic-resistant diseases. The Swedish meat industry foresaw a crisis of confidence among its consuming public. It wanted to get ahead of the issue and to keep a level playing field for everyone raising livestock and poultry. The meat industry itself lobbied the legislature for a law forbidding the continuous feeding of antibiotics to animals raised for food. With the support of the meat industry, the law passed in 1986, making Sweden the first country in Europe to pass such a law. By 2006, a similar law went into effect for the entire European Union.

If you are part of the livestock or meat industry you have two choices: You can acknowledge legitimate issues and work to fix them, or you can engage in blanket denials. My hope, as part of the cattle and beef industry, is that the legitimate concerns will be recognized and solutions will be adopted by the industry itself. There is no better way to silence the critics.

Here are the things I think need work:

1. **Better grazing management.** Too much grazing is occurring without good planning and oversight. Well-managed grazing is environmentally essential; poorly managed grazing is damaging. Every farmer and rancher needs to get on board.

2. **Stop killing primary predators.** Predators are essential to ecosystems.[7] Ranchers need well-functioning ecosystems even more than the rest of society. We must learn to deter predation and co-exist with these essential creatures.

3. **Stop feeding drugs and other junk.** A shocking list of drugs, including antibiotics and beta-agonists, and industrial by-products of all kinds are regularly fed to cattle, mostly at feedlots. This creates unhealthy animals, results in foods that may be unsafe for humans, and contaminates the waste stream. Nothing should be fed to cattle other than pure feeds, comparable to what they'd eat in nature. Although I do not believe grains are inherently bad for cattle, I do believe they should be limited because feeding them to cattle is a poor use of resources and leads to additional water and atmospheric pollution. All types and classes of cattle should be foraging to the greatest extent possible.

4. **Stop using hormones.** No growth hormones should be used on dairy or beef cattle, period. As with beta-agonists, it creates health and welfare concerns for cattle and leads to food that may be unsafe for humans. It also limits the markets for U.S. beef. The European Union banned the practice in 1981. All use of growth hormones should immediately stop.

5. **Stop putting calves in feedlots.** I believe the optimal way to raise cattle is on grass for their entire lives. To the extent that I find

feedlots acceptable, I do not find them acceptable for young cattle. Health and welfare problems are magnified for young cattle.[8] Cattle should not go into a feedlot until they are at least a year old, preferable 18 months.

6. **Stop slaughtering young cattle.** The routine slaughtering of young cattle (less than two years of age) is a relatively new practice in the United States. It has been abetted by the use of beta-agonists, hormones, and high-concentrate feeds, all of which are practices that should be stopped (or, in the case of feeding concentrates, minimized). Cattle should be raised to full maturity before going to slaughter (a minimum of two years of age). This is a better use of resources and ensures better meat.

7. **Stop long-distance transport.** Cattle do not lie down in truck transport. When they do, they are trampled to death. This is why it is essential that cattle be shipped only moderate distances. According to Animal Welfare Approved standards, cattle transport should never exceed eight hours.[9] Shipping cattle long distances where they must stand longer is inhumane.

8. **Improve slaughter practices.** Cattle should be humanely handled at slaughter, period. It is not only the right thing to do, it has a strong correlation to meat quality and safety. The best way to ensure appropriate, low-stress handling is for the people who own the animals, and are familiar to them, to do it themselves. This goes a long way toward ensuring the animals remain calm. Additionally, video cameras should be installed at all slaughterhouses to ensure humane handling of all animals at all times.

Reading this list in isolation, you might conclude that I am a beef detractor. Instead, I consider myself a concerned member of the cattle industry. It is my hope that everyone who reads this list and who is involved with the raising, transporting, or slaughter of cattle will instead look at this list as a *call to action*. If we wish to continue to exist, we can and should do better.

WHY EAT ANIMALS?

IN ADDITION TO WHAT'S ALREADY BEEN EXPLORED in this book, two major ethical questions surrounding beef consumption remain. One is whether it's morally acceptable to eat meat at all. The other is whether eating meat aggravates world hunger. I will address the first momentarily, and start with the latter. The idea undergirded the hugely popular and influential book *Diet for a Small Planet*, which argued, essentially, that the world's finite resources are stretched thin and would quickly be expended if a growing population of humans continues eating meat. The raising of livestock, especially cattle, the argument goes, is uniquely resource-intensive and cannot be morally justified in a world where (now) some 900 million people don't have enough to eat. In various forms, we continue to see this line of reasoning everywhere today in the materials of vegan and environmental groups.

I agree that meat and dairy consumption is out of balance: There's more than necessary in industrialized countries and not enough in developing countries, where malnutrition generally, and deficiencies of protein, iron, and vitamin B_{12} specifically, are rampant. But there are so many things wrong with the assertion that eating meat adds to global malnutrition and starvation that it's hard to know where to begin.

Perhaps first I will point out that it can only be sensible to *quit* eating meat for this reason if doing so actually *aids* in relieving global hunger. For me, as a lawyer, this is so self-evident it should hardly need to be stated. And yet, in over 20 years of reading various forms of the "livestock aggravates world hunger" argument, I have never seen anyone effectively demonstrate that if you *stop* eating meat you will *help* world hunger. Rarely is such proof even attempted.

When distilled down to its essence, this is not really an argument that by refraining from eating meat you will help feed others. Instead it's more an endorsement of a principle of food equity: that it's *unfair* to eat resource-intensive foods while others have insufficient food. But we could just as easily argue that we should refuse to drive because billions of people in the world cannot afford cars; we should refuse to use air-conditioning; we should refuse to take airplanes. I don't see how doing any of those things helps a single person in need, therefore I find none of those arguments compelling.

Global hunger is not actually, and has not at any time in recent decades, been a product of an inadequate world food supply. Food and nutrition expert Dr. Marion Nestle points out that the global food system produces some 3,800 calories per day for every man, woman, and child on earth, which, she points out, is almost double what's necessary for adequate nourishment.[1] For the last four decades, per capita food production has actually grown at a pace 16 percent faster than the world's population.[2] In his book *World Food Security*, Dr. Martin M. McLaughlin, who has worked on food security issues and taught university courses on the subject for decades, makes plain that world hunger has very little connection with the quantity of food the globe produces.[3] Poverty, not food shortage, is the key, says McLaughlin. "Hunger . . . is a political and social problem," he writes. "It is a problem of access to food supplies, of distribution, and of entitlement."[4]

Moreover, livestock farming is not the province of the rich—in fact, very far from it. It actually helps more poor people than it hurts. A recent article in the journal *Nature* points out that one billion of the world's poorest people depend on livestock for their survival.[5] Likewise, a 2013 FAO report states: "Hundreds of millions of pastoralists and smallholders depend on livestock for their daily survival and extra income and food."[6] In many developing countries, many poor families, including those who own no land, have a cow or goat or some chickens, and the eggs, milk, and meat make up an irreplaceable component of their income and food. "Almost every smallholder farming family in a developing country owns livestock, whether chickens, rabbits, sheep, goats, pigs, cows, buffaloes, donkeys, horses, yaks, llamas, or camels," states a 2014 article by an Indian agriculture official.[7] "Livestock development benefits poor rural families, many of them engaged in farming but not owning land."

Livestock keeping offers numerous salient advantages in gaining food and financial security not afforded by plant crops. In contrast with crop farming, which produces sporadic, seasonal, perishable products, livestock is an asset that can be maintained for short or long periods of time then quickly converted to food or cash when needed. This has been the case since people began keeping farm animals, Simon Fairlie points out. "[A] main role of animal husbandry has been to provide food security: 'The purpose of domestication was to secure animal protein reserves and to have animals serve as living food conserves.'"[8] This is why livestock are sometimes referred to as "an ATM for poor farmers." The world over, the flexible nature of animal keeping has always been among its primary benefits, Harold McGee points out: "Livestock not only transformed inedible grass and scraps into nutritious meat, but constituted a walking larder, a store of concentrated nourishment that could be harvested whenever it was needed."[9] Additionally, in many parts of the world farm animals raised for meat and milk also provide invaluable labor and transportation services. Oxen still pull plows and carts on a large portion of the globe. These are all attributes utterly unique to animal keeping.

Women, who make up the majority of animal tenders in the developing world, are often livestock's greatest beneficiaries.[10] Animals provide women reliable income and protein-rich foods for their own families, both available on an as-needed basis.

An article in the science journal *Nature* points out, as well, that crop and livestock farming are highly complementary. "Half the world's food comes from farms that raise both. Animals pull ploughs and carts, and their manure fertilizes crops, which supply post-harvest residues to livestock."[11] In his book *Feeding People Is Easy,* veteran British science reporter Colin Tudge, who has traveled the world extensively and reported for decades on food and agriculture, states that "pastoral farming is very important indeed," declaring: "The oft-bruited generalization—that we could most easily feed the world if everyone was vegetarian—is simply not true." Among the reasons he points to are that "there is no system of all-plant agriculture that could not be made more efficient, in biological terms, by adding in a few livestock, provided they are the right kind, and are kept in the right numbers, in the right ways."[12]

That farm animals are the lifeblood for hundreds of millions of the world's poor became much clearer to me after I attended an international convening of smaller-scale livestock keepers. The Livestock Futures Conference in Bonn, Germany, hosted by the nonprofit League for Pastoral Peoples, gathered 70 livestock keepers and researchers from 16 different countries and several continents. Among those in attendance were people involved in camel herding from Pakistan, cattle herding from Uganda, sheep herding from Germany, and goat herding from Argentina. (Henning Steinfeld, lead author of the FAO's *Livestock's Long Shadow* report, was there, too.) I was honored to have been invited to speak about issues facing livestock farmers and ranchers in the United States. The organization's founder, a remarkable German woman named Dr. Ilse Köhler-Rollefson, a doctor of anthropology and veterinary medicine who has lived for extended periods among camel-herding people in India, has labored for years to raise the profile of smaller-scale livestock keepers before the world's policy organizations. On the global level, just as on a national one, the biggest players tend to be given greatest consideration.

The Livestock Futures Conference highlighted the environmental and societal damage caused by industrialized livestock production, contrasting it with the social and ecological contributions of smaller-scale farmers and herders. Research and testimony described wide-ranging benefits to smaller keepers, including environmental and climate protection, cultural preservation, tourism, and support of local labor markets. Presenters showed that in many of the globe's arid regions, in places where land cannot be put to other uses, the small-scale livestock sector is responsible for the largest share of animal production, all told making up 30 percent of world production of animal-based foods. Despite the wealth of measurable contributions of small-scale keepers to food security, they are still being ignored in national and international policy.[13] The conference was part of a multiyear strategy by the league and its allies to change that.

Even if we accept without proof that eliminating livestock would lead to a greater supply of food for the poor (which I definitely do not), the notion relies on exceedingly fuzzy math. Let's look at some of the problems. First off, it assumes that grain currently grown and fed to livestock would still be grown and would somehow end up being made available to the world's poor. This completely ignores the realities on the ground. If livestock were no longer generating a demand for grain, why would farmers and agribusiness companies continue growing it? They would not. Like anyone engaged in the production of any commodity, they would adjust, and farmers would shift to growing crops for which there was the most demand. More sugar beets, perhaps? In the alternative, they might convert the land to other, non-agricultural uses. Under either scenario, global hunger is in no way reduced.

For another thing, the nutritional value of the world's supply of meat and milk would be difficult to replace with foods from plants, especially for people in the developing world and children. To substitute for all animal-based foods would take far more than a one-to-one pound-for-pound or calorie-for-calorie replacement by grains and soy. The Pimentels point out: "Animal proteins contain the eight essential amino acids in optimal amounts and in forms utilizable by humans for protein synthesis. For this reason, animal proteins are considered high quality proteins. By comparison, plant proteins contain lesser amounts of some of the essential amino acids and are judged to be lower in nutritional quality than animal sources."[14] This is especially important for children, they note, whose rapidly developing bodies particularly benefit from nutrient-dense foods. "Another advantage of animal products over plant products as food for humans, especially children, is the greater concentration of food energy per unit of weight compared with plant material. For example, . . . beef has three times as much food energy per unit of weight as sweet corn."[15] Nutritionally, animal-based foods are more important to the world's poor than other foods, and for the one billion of them who raise livestock, it helps them feed themselves. Eliminating animals from the food system would likely make the world's hungry *more* food-insecure, not less, and more dependent on government assistance.

Moreover, while world grain use for livestock is significant (and I believe it's too high), it's far less than what people generally assume. Cattle in the developing world are usually fed little or no grain. In the United States, the breeding herds of beef cattle (about 30 million animals) are generally also maintained without grain. Nearly all steers and heifers raised for beef in the United States are raised on mother's milk and pasture, then fed grains only in the latter portion of their lives. And overall, even for cattle fed grain in industrialized countries (both beef and dairy types), a large portion of their diet still comes from forage or farm by-products (like straw or rice bran). An article in *Nature* noted that around 70 percent of grains used by developed countries are fed to animals, with 40 percent of such feed going to ruminants, mainly cattle.[16] As I have argued several times in this book, this amount can and should be pushed down significantly. Even so, the same article points out, much livestock feed in developed countries comprises plant matter inedible to humans. "Even where large quantities of cereals are consumed by ruminants, up to 60% of their diet comes from high-fibre feed that humans cannot digest . . . In the European Union, more than 95% of milk comes from animals fed on grass, hay and silage, supplemented with cereals."[17]

Moreover, even in the developed world, some cattle milk and beef comes from animals raised entirely with very minimal or no grains. Some American beef cattle, our own among them, are raised from birth to slaughter with no grain. "Cattle in New Zealand's exemplary dairy industry obtain 90% of their overall nutrition by grazing pasture." New Zealand's dairy industry and grass-fed cattle ranchers in many parts of the world, including the United States, demonstrate the feasibility of a worldwide transition back toward forage-based diets for ruminants, including cattle. Some production would be lost, but that would be more than offset by the overall benefits to the environment, animal health and welfare, and human health.

Energy use for cattle fed from their own foraging is so negligible that beef produced in this way is actually less energy-intensive than grain production. The Pimentels point out in *Food, Energy, and Society* that whereas crop cultivation adds significantly to the energy use of grain-fed livestock systems, raising cattle on grass takes little energy. "[I]n contrast [with grain-fed], cattle grazed on pastures use considerably less energy than grain-fed cattle."[18] The textbook quantifies energy inputs for grass-fed beef compared with grain-fed as follows: "Current yield of beef protein from productive pastures is about 66 kilograms per hectare, while the energy input per kilogram of animal protein produced is 3,500 kilocalories. Therefore, animal protein production on good pastures is less expensive in terms of fossil energy inputs than grain protein production."[19]

The other part of the fuzzy math problem is about land where cattle and other livestock are currently grazing. The "stop eating meat to reduce world hunger" notion assumes that if livestock disappeared, a significant portion of the land where they graze could (and would) be used to raise food for humans instead. This is wrong on several levels. First, just as there's no reason to believe grain production would continue without livestock generating a demand for it, there's no reason to believe land currently used for grazing would be used to grow food for the world's poor. Whoever owns or controls the land would find other, more profitable uses. In the alternative (such as federal lands grazed by U.S. cattle), the land would simply cease to be used for any food production. Thus, removing livestock would not free up land for plant-based food production, as people making this argument often assume.

Second, and this point is critical, the vast majority of the world's grazing takes place on land that *cannot* be used to grow crops. As David Montgomery succinctly states: "Sheep and cattle turn parts of plants we can't eat into milk and meat."[20] *Food, Energy, and Society* notes the prevalence around the world of livestock raised on "free energy sources." These include forage growing along paths and other "interstitial spaces" that would not be used for crops or other purposes, and straw left after harvest of rice or similar grain crops, which can be fed to animals.[21]

Word

The textbook *Soil and Water Conservation* defines *rangelands* as "soil on which the native vegetation is predominantly grasses, grass-like plants, forbs or shrubs suitable for grazing or browsing." It notes that "[n]early half of the land on Earth can be classed as rangeland," and says, "Most of it is either unsuitable or of low quality for use as tilled cropland because it includes steep areas, shallow and/or stony soils, or dry and/or cold climates."[22]

Likewise, the National Sustainable Agriculture Information Service provides the following explanation of the unique role of grasslands and ruminants like cattle in the global food system:

> [Grassland] ecosystems are naturally able to capture sunlight and convert it into food energy for plants. . . . [M]ost of the land in the U.S., and indeed in most countries of the world, is not tillable and is considered rangeland, forest, or desert. These ecosystems can be very productive from a plant bio-mass perspective, but since they are generally non-farmable, the plants they produce (grasses, forbs, shrubs, trees) are not readily usable (from a digestive standpoint) by humans.

> *However, grassland ecosystems (both rangeland and temperate grasslands) produce plant materials that are highly digestible to ruminant animals. . . . Grazing of native and introduced forages on grasslands and rangeland thus is a very efficient way of converting otherwise non-digestible energy into forms available for human use: milk, meat, wool and other fibers, and hide.*[23]

This point is so important it's worth stressing again: This miraculous transformation of sunlight into human food via grazing animals is mostly occurring in areas that cannot be used to grow crops. This crucial fact is nearly always ignored by (or perhaps unknown to) beef's critics. Colin Tudge uses his book *Feeding People Is Easy* to raise awareness:

> *In many parts of the world, at least in some seasons, it is very difficult to raise crops at all. Arable is all but impossible when the land is too high, steep, cold, or wet or if it rains too much in the season when the grain should be ripening . . . [A]nimals of one kind or another muddle through anywhere—living as camels and goats may do on the most meager of leaves that poke through between the thorns of desert trees, or as reindeer do on lichen, or as long-wooled sheep and shaggy cattle do in British hills on the coarse grasses that grow between the heather; and in times of drought or the depths of winter there may be nothing to eat at all except for beasts that fattened in better times.*[24]

A recent grasslands university textbook reinforces Tudge's observations with respect to rangelands, stating that "[b]ecause they typically experience low and unpredictable rainfall and often have associated low soil fertility, rangelands generally cannot sustain crop agriculture without irrigation."[25] Making use of these lands is and has always been the very special place for the world's herbivores, the text then notes. They convert "low quality fibrous plants into products such as meat, milk, and blood that humans can readily digest." For this very reason, it continues, "harvesting products from herbivores has been a defining element in the relationship between humans and rangelands worldwide for millennia."[26]

One of many world examples of people using livestock as rangeland converters are the Dodos of northeast Uganda. They feed their cattle no grain, only pasture forage unsuitable for human consumption, and raise them without fossil fuels. The Dodo tribe illustrates the crucial and versatile role livestock can play for humans. *Food, Energy, and Society* summarizes the benefits: "First, the livestock effectively convert forage growing in the marginal habitat into food suitable for humans. Second, herds serve as stored food resources. Third, the cattle can be traded for sorghum grain [for human consumption] during years of inadequate rainfall and poor crop yields."[27] These are precisely the properties that make livestock irreplaceable to people throughout the world.

In the United States, since long before the arrival of humans, a large portion of ground, especially in the West, has been unsuitable for crop cultivation. Some of this land is arid or semi-arid; rains may be insufficient or fall only at the wrong time of year for crop cultivation, or its topography is too hilly, or too rocky. Having resided for the past 12 years in Northern Coastal California on land where crops cannot be grown, I grasp such limitations much better than I once did. I understand implicitly how windiness, dry, cool summers, and steep, rough terrain are all conditions that are fine, even ideal, for grass and livestock, but render crop growing impossible.

"The [U.S.] pastureland and rangeland are marginal in terms of productivity because there is too little rainfall for crop production," note the Pimentels.[28] These areas are where the vast majority of America's cattle are located. According to the U.S. Beef Board, 85 percent of the land grazed by cattle in the United States is land that cannot be farmed.[29] This precise number, since it comes from the beef industry itself, obviously should be taken with a grain of salt. But it suggests that cattle grazing in the United States is largely occurring on non-farmable rangelands. According to a highly credible and impartial source, a recent university textbook, in California, 57 million acres, almost 60 percent of the land, is characterized as rangeland, about 34 million acres of which is actually grazed.[30]

Yet in California, as elsewhere in the United States, rangelands continue to be chipped away by more intensive land uses. Foothill rangelands, especially, are being converted to wine-grape growing, housing, and urban developments. The grasslands textbook states: "[T]hroughout most of California, range has given way to other, generally higher value but also more intensive, land uses."[31]

However, the same textbook holds out hope for a growing recognition of rangelands' societal and ecological value:

> As range ecosystem services other than livestock production become increasingly valued by society, the additional benefits that we derive directly from primary production and the soil system have gained greater recognition: provision of irrigation and drinking water, recreational opportunities, wildlife habitat, open space/viewshed, rural lifestyle, biodiversity, and carbon storage. It might even be argued that the most important role that primary consumers play in 21st century California is not providing the traditional livestock products of meat, milk, and fiber, but rather acting as a bulwark against the conversion of range into housing developments, vineyards, and other more intensive land-uses that do not provide the multiple ecosystem services bestowed by range ecosystems.[32]

The presence of grazing animals on such non-farmable lands enriches the world's food supply regardless of the efficiency with which the animal converts the feed to flesh and milk. The oft-quoted statistics about the "inefficiency" of cattle converting feed to flesh are irrelevant. "However efficient the conversion ratio of any given animal may be," Simon Fairlie points out, "if it is grazing entirely on land which could not otherwise be used for arable production or some other highly productive activity, then it cannot be said to be detracting from the sum quantity of nutrients available to the people of the world, but adding to them."[33] Moreover, he points out, where animals graze on land unsuitable for crop cultivation, they are "relieving pressure on arable land, and helping to retrieve otherwise inaccessible nutrients and bring them within the food chain."[34] To this I would add livestock grazed on cover crops or fed farm by-products, as is the case for a large number of the world's farm animals. These methods, too, create food for humans using only feed sources that are not directly usable as human nourishment.

I hope this discussion puts at ease the mind of any reader who has hesitated to eat beef based on concerns about world hunger. In the majority of the world, cattle are fed little or no grain and are raised mostly on non-farmable lands. For Americans and other people in the developed world, where grain is used as part of cattle feed, we have the choice to seek out and buy beef and dairy products from animals raised on forages rather than grains and soy. As described in earlier parts of this book, there are human health and animal health and welfare reasons to do so, and by choosing grass-based foods we help maintain our nation's grasslands, which are the most environmentally beneficial of all lands used for agriculture.

The other aspect of the moral question about eating beef is whether it's acceptable for humans to eat meat at all. In answering this, I'd like to rely less on data and statistics and bring some of my own personal experiences to bear. It's been 14 years since I began working on farm-related issues for Waterkeeper. Although the job was focused on addressing pollution, to me the animals were equally important. While other environmental groups were publicly advocating addressing problems from industrialization with more effective waste containment or treatment, I found those approaches far too narrow. They ignored factory farming's greatest evil: animal cruelty. Even worse, by endorsing steps geared toward pollution reduction that failed to improve farm animals' lives, they were further entrenching the current system.

Fortunately, my boss felt the same way. Bobby Kennedy Jr. has cared passionately since childhood about every creature from sow bugs to blue whales. He heartily endorsed my advocating for farming that was ecologically sound *and* provided animals good lives. Having closely viewed the brutality of industrial production, we felt morally obligated to seek improvements in farm animal welfare.

Over the years, my writings and speeches often addressed the ethics of *how* meat was produced, but until a few years ago they never spoke to whether or not it is ethical to eat meat at all. Then I was invited by an ecological journal to write an essay arguing that an environmentalist need not refrain from meat eating. Following publication of the essay, I was extended invitations to participate in several live debates about the ethics of meat eating, two of which I accepted. Both times (somewhat ironically, since I am still a vegetarian) I represented the "pro meat" position. I later wrote a piece for TheAtlantic.com titled, "Can Meat Eaters Also Be Environmentalists?"[35] Even though I never sought to focus my energies on the question of whether people should eat meat, it seemed important to refute the increasingly prevalent notion that people who cared about the environment should avoid it. I was becoming a de facto vegetarian meat advocate.

My first piece, titled "Animals Are Essential to Sustainable Food," proposed that today's debate over meat is characterized by polarizing, oversimplified rhetoric, pitting an implacable, defensive agribusiness in one corner against equally intractable vegan activists for abolition of all animal farming on the other. "[The abolitionist vegans'] fervent advocacy echoes prohibitionists at the dawn of the twentieth century," I wrote, "some of whom attacked apple trees with axes because they were the source of hard cider. Like the prohibitionists, activists against meat are fueled by the excesses of the day."[36] Factory farms, not animal farming, are the real problem, I urged. And industrial methods have fostered a growing disillusionment with the meat industry among broad swaths of the American public, well beyond vegan and vegetarian circles.

I will be the first to agree that industrial methods for raising farm animals are indefensible, and I believe all people should join in rejecting them. Having seen it in all its gory details, I have no qualms about calling industrialized animal production a routinized form of animal torture. While Prohibitionists attacking innocent apple trees with axes seem absurd to us today, a lot of discussion over the ethics of meat eating likewise focuses on the wrong villain. Industrial animal production is rightly vilified; animal farming, on the other hand, is not.

What has really fostered my interest in the debate over meat eating is not a desire to encourage meat consumption but a longing for some nuance in the discussion. The issue is far from black-and-white, and polarized camps lobbing accusations at each other only hinder movement toward a better system. Building a food system that is more ecological and more humane is far more important to me than whether or not so-and-so is eating meat.

I believe the real issue is whether we humans are living up to our responsibilities of good stewardship of animals and the earth. Michael Pollan and others have proposed the idea that animals "chose" domestication based on a sort of "bargain" with humanity. I put the words *chose* and *bargain* in quotes because, quite clearly, no individual wild animal made a conscious decision that its species should become domesticated. Instead, domestication likely happened gradually over many generations as some animals found advantages to having a certain amount of human contact. Humans "agreed" (again, in quotes, because the bargain was entirely implicit) to provide essentials to animals—food, shelter, and protection from predators being foremost among them—in exchange for the animals providing humans food in the form of eggs, milk, and meat. (With dogs the terms of the bargain were different: For being provided protection and nourishment, dogs exchanged assistance in hunting, early warning, and self-defense).[37] However, it's reasonable to assume, as well, that animals would never have opted for such an arrangement if torture had been part of the deal. Stated simply: By raising animals in factory farms, humans are violating their age-old contract with domesticated animals.

With those raised for food, the very idea that the individual animal's dignity matters seems to have been abandoned in the United States sometime around the mid-20th century. Just as confinement animal operations were becoming the norm, every agricultural college in America changed the appellation of its "animal husbandry department" to "animal science department," which is emblematic of the shift in mind-set.

Agribusiness has long defended its methods by pointing to their prevalence. But we all know that just because something is widespread doesn't mean it's acceptable, let alone right. Factory farms are undeniably inhumane. The worst practices are narrow metal cages for pregnant sows, wooden crates for veal calves, and wire cages for egg-laying hens. But beyond that, the everyday workings of industrial facilities utterly fail to provide animals decent lives. Continually keeping animals in foul-smelling cramped conditions, depriving them of all pleasures and basic necessities like exercise, fresh air, sunshine, and a soft place to lie down, cannot be called humane. Whatever rationale is offered for these practices—"efficiency," "cost of production," "affordable food," "feeding the world"—these systems remain morally indefensible.

Grazing animals, especially those raised for meat (rather than milk), have fared better than others. Their unique capacity to sustain themselves on grass has been their saving grace—keeping them outdoors on growing vegetation is often the most economical way to raise them. Nearly all cattle, including dairy heifers, spend their early life on grass. Once mature, most (although not all) dairy cows with modern genetics, bred for high-volume milk production, are confined and fed concentrates, which is the only way to achieve their genetic potential for milk production. Their time on grass is then over.

Beef cattle have it better. Those raised for meat (not kept for breeding) typically go to a feedlot sometime before one year of age. Even there, they are out in the open air and have the benefit of soft ground for lying and standing. The breeding beef animals (mother cow herds and bulls) are the most fortunate. Generally, they spend their entire lives on pastures or rangeland, having a daily existence not unlike that of their wild ancestors. Because a cattle ranch's success depends on mother cows being able to survive and give birth without human assistance, beef cattle have long been selected for heartiness and good calving ability. In this, the interests of the animal and those of the rancher perfectly coincide. These traits help rather than hamper the quality of life experienced by the individual animal.

In fact, beef cattle are so much better off than other animals raised for food that, as I mentioned earlier, my Humane Society friend says he considers it better to eat beef than eggs. Soon after my work became focused on agriculture I, too, became persuaded that cattle raised for beef are the luckiest of all farm animals. And from an eater's perspective, I'd much rather my food source had spent its life exercising, breathing fresh air, and grazing meadows than cooped up in a crowded, stinking metal warehouse. Why would I want to eat food that originated from a place I would never want to visit?

On top of those issues, there is the amount of meat per animal to be considered. Previously I noted that taking the life of a single steer provides more than ample meat for two families for an entire year, an industry average of 475 pounds of beef per carcass. In stark contrast, the industry average weight for a whole chicken carcass is 4 pounds, with about 70 percent of that being meat; a chicken will thus yield less than 3 pounds of meat. You have to kill more than 150 chickens to get as much meat as you receive from one steer. It's hard to know how to compare the morality of killing 150 chickens versus killing one steer. But for me, it was among the reasons I began favoring beef over other animal-based foods even before first stepping foot on my husband's ranch.

I've never adopted the view that eating meat is inherently wrong. We can debate until the cows come home about evolution of teeth, digestive tracts, and other aspects of human physiology to make a case either that humans are "intended" to eat meat or not.[38] But to my biology-trained brain, those points were never very persuasive. Humans may have evolved from herbivores if you go back to a certain moment in time. But the ancestors of those animals were omnivores and carnivores. And it's now believed human ancestors began eating meat at least 2.6 million years ago, with a major uptick in meat consumption occurring around 1.5 million years ago. Clearly, a great deal of evolving has taken place in those millions of years. Any way you slice it, modern humans come from a very long line of meat eaters.

My view boils down to this: Humans are animals belonging to a food web. That web includes animals eating plants, animals eating other animals, and even plants eating animals. As this book has been at pains to portray, all life starts from the earth and returns to the earth, the bodies of all plants and animals nourishing future generations of plants and animals in an endless cycle of regeneration. Ashes to ashes and dust to dust. To me, something so fundamental to the functioning of nature cannot be regarded as morally problematic.

As I grew more aware of how animals in the food system live, it also brought me to the realization that, morally speaking, there is little difference between eating meat and refraining from meat if you are still eating dairy and eggs—in other words, practicing classic vegetarianism. Dairy cows all become beef eventually and, in many ways, their lives are not nearly as good as cattle raised for beef. Most laying hens spend their lives crammed into wire cages (so-called battery cages), then are usually unceremoniously vacuumed up for use in things like canned chicken soup and pet food. To feel any sort of moral superiority for not eating beef while eating dairy products and eggs seems absurd. Once that became clear to me, it bridged any ethical distance I might have once felt between myself and my husband's vocation.

Still, I initially had some uneasiness about moving to a cattle ranch. Although I supported it *in principle,* I wasn't sure I'd be comfortable living day after day in the midst of an active operation. Would I find it upsetting to be surrounded by animals I knew would one day be sent to slaughter? And even more to the point, would it make me feel guilty to know that our living came from their deaths? I figured I'd keep myself at arm's length to avoid any potential discomfort.

What happened instead was just the opposite. For exercise and to take in the area's natural beauty, I took long walks nearly every day through our land. Almost by accident, I began regularly spending time in the company of our cattle. I saw our mother cows ambling as they ate, socializing with their sister herd members, congregating around the water trough, calling their babies, licking one another's necks. I watched calves frolicking in the grass, racing around after one another at twilight, running to their mothers for a long, warm drink of milk. I saw the bulls and cows nuzzling each other in courtship before and after mating. It was easy to see that the lives of these animals were well worth living. The more I meandered our meadows, the more I sat on our fences observing, the more I valued and appreciated what was happening in my midst. Everywhere I looked I saw animals living well and well-cared-for land.

After a few months, I told Bill I wanted to learn to do everything on the ranch. This surprised him a bit (he, too, had expected his vegetarian wife to want to keep some distance from ranching operations), but he readily agreed. Soon I became the wide-eyed, unskilled, displaced city-dweller ranch hand, helping him and our ranch manager with whatever needed to be done each day. Occasionally, this involved fixing a fence or a water trough (although my role was usually tantamount to holding the tools). In the dry season, it meant bringing our cattle some hay. Daily, it meant taking my pocket-sized notebook and walking or riding on horseback through the herd, making sure each animal was healthy and accounted for. Over time, I came to know each herd member individually, learned some things, and was given more responsibility.

Eventually, and for several years following, I became the primary labor on our ranch. I loved doing physical work, outside, being among the animals, and the challenge of problem solving, the constant companion of every practicing farmer and rancher. I especially enjoyed being useful to the animals, like when I'd reunite a mother with her new calf who'd slid beneath a fence, or when we'd successfully graft a spare twin calf onto a mother whose calf was stillborn.

A lot of my time was spent watching and interacting with wildlife, as well. I became daily witness to nature's beauty, its force, and its cycles. A mother bobcat stealthily stalking her prey; an osprey cruising silently overhead with a fish clasped in its talons; a wake of vultures feasting hungrily on a deer carcass. Grasses and wildflowers sprouting, blooming, drying, dropping their seeds, dying back. Life-giving rains coming and the cycle beginning anew.

For me, these experiences reinforced that all life is connected. Living in Manhattan, as I had for nearly five years just before moving to the ranch, it had been easy to see myself individually, and humanity collectively, as isolated from the rest of the natural world. Working every day out on the ranch, maintaining such a view was impossible.

It became increasingly plain to me, as well, that even the most conscientious agriculture is a major disturbance. No matter how well done, it will invariably have profound impacts on wild creatures and plants, soils, and water. Our expectations of farmers and ranchers should not be zero environmental impact, which is unattainable. The goal, instead, should be food production done as harmoniously as possible with natural elements existing in the wild.

Now consider crop cultivation. Especially if done using typical farm machinery, but even with non-motorized plows, it represents an enormous interruption to naturally occurring vegetation and animals. Plows scrape, cut, and chop up the earth's surface and whatever is growing there, tearing up complex communities of everything from symbiotic microorganisms to rabbits and snakes. In the developed world most plows are dragged behind enormous, heavy tractors, which is followed later in the season by huge harvesters that crush and shred every plant and animal in their path. These machines are Armageddon for billions of soil-dwelling creatures along with every form of wildlife that resides in or on the ground. There's no avoiding the reality that crop farming is the most disruptive of all agricultural acts.

At the start of his beautifully photographed and eloquently written *The River Cottage Meat Book*, British chef Hugh Fearnley-Whittingstall poses the questions: "Why do we eat meat? And is it right, morally, that we do?"[39] In answering, he starts by noting that no matter what we do, it's a fallacy "to maintain that we can live in complete harmony with the rest of animal kind." The actions of all animals, he points out, whether intentionally or not, will affect others: "The undeniable fact is that any species' pursuit of its interests will always have an impact on the rest of the planet's life—the fox impacting on the chicken population, the flea on the cat, the beaver on the forest, and the sheep on the grass."[40]

I agree with Fearnley-Whittingstall's implicit suggestion that no human endeavors, and certainly not farming, can be done without enormous impacts on other life-forms. The notion that we can "eat cruelty-free" by avoiding foods derived from animals is nonsensical if we really look at agriculture, especially crop farming used to produce plant-based foods. Cultivations of soy, grains, fruits, and vegetables are all highly altering of habitat and have immediate and ripple effects on literally billions and billions of creatures of all types.

What I'm looking for is agriculture that respects all life and follows nature's model. Answering the question: *Am I eating food derived from an animal?* tells you very little about the impact production of that food has had on nearby animals and plants. All farming, and especially crop farming, necessarily kills a lot of animals of all shapes and sizes. The more meaningful question is: *Has this food been produced as nature functions?* And for me, it is clear such farming embraces animals.

The great Sir Albert Howard, a godfather of modern organic farming, viewed animals as inextricably linked to ecologically sound food production, calling them "our farming partner." Howard said: "In Nature animals and plants lead an interlocked existence. The connection could not be closer, more permanent, or more crucial. We can observe this partnership in operation in the forest, in the prairie, in marshes, streams, rivers, lakes, and the ocean."[41] Howard's point, I believe, is that nature's way of converting water and sunlight to energy, in the form of food, is a complex, multi-stage process in which no part exists in isolation. Each component—every ray of sunshine, every drop of water, every clump of soil, every plant, every insect, every grazing animal—has many and varied roles and effects. The more farming systems reflect such complexity, the more they are ecological.

One of the most thorough and thoughtful explorations I've seen of why we should farm with animals is in Simon Fairlie's book *Meat: A Benign Extravagance*. Fairlie's perspective is informed by his diverse experiences, which include environmental writer, farmer, and former vegetarian. He believes farming ecologically involves animals. His book describes an optimal food system where cattle and other grazing animals convert non-tillable lands into contributors to the food system and omnivorous animals like pigs and chickens make good use of farm by-products and food scraps. He also discusses at length the value of manure for post-fossil-fuel fertility.

Additionally, Fairlie points out that "wherever there is livestock there is the opportunity to garner and concentrate fertility from the wilder environment . . ."[42] This is especially the case for grazing animals. More efficiently than machines, Fairlie notes, grazing animals "move nutrients from where they are not needed to where they are required." This is particularly true for phosphorus, which, unlike nitrogen, cannot be obtained from the air. "[P]lants cannot extract phosphorus from the atmosphere, so the role of animals in importing surplus phosphorus from outlying areas could be crucial," he argues.[43]

Fairlie also points out that domesticated grazing animals have taken the place of wild herbivores in keeping a balance of open spaces with wooded areas or, as he puts it, light with shade. The balance, he argues, would massively shift without the presence of grazing animals:

> *Only livestock can engineer the balance that any society seeks between the realm of light and the realm of shade on any scale beyond the arable. Even in a full-blooded fossil fuel economy, JCBs, timber harvesters and other wheeled monsters are fighting a losing battle with nature unless they enlist the help of quadrupeds. Where livestock are allowed*

*to roam they bring grass, and where they are excluded trees
grow, and it is a relatively effortless matter for humans to
calibrate their performance to our will or whim.*[44]

In a related vein, the university textbook *Soil and Water
Conservation* describes the value of livestock for managing fire
risks on a broad, landscape scale. "When livestock graze these
fire lanes, forest roads, and forestlands in general, they reduce
the fire hazard by removing vegetation that is flammable when
dry."[45] The way that cattle manage vegetation, holding back
the spread of woody plants and keeping open spaces open, is
something few of us pause to consider or appreciate. The look
and functionality of our landscapes would be radically different
without them, and in many ways for the worse. In short, humanity needs grazing animals.

We hear a great deal about the planet becoming crowded and
harder to feed. All too often we hear that livestock are part of the
problem. Because it doesn't fit neatly into the advocacy narrative
for vegans and environmentalists, though, we rarely hear about
crop cultivation destroying agricultural land, which is actually the
greatest threat to humans' ability to feed themselves in the future.
In *Dirt: The Erosion of Civilizations*, David Montgomery tells the
history of societies failing to properly steward their soils, to the
point where their lands are no longer usable for crop cultivation;
and this continues at an alarming pace today. *Food, Energy, and
Society* notes that, worldwide, more than 50 million acres of agricultural land is abandoned annually because of soil erosion and
salinization from irrigating crops. "During the past 40 years, about
30% of total world arable land has been abandoned because it is
no longer productive," the book notes. It is estimated that about
half of the land currently under cultivation will be unsuitable for
food production by the middle of the 21st century.[46]

Individuals and groups are rightly concerned about adequate
food supplies for the future. But they would do well to focus their
attention on this imminent crisis, and on the way livestock are
managed on the land, rather than on the absolute number of

livestock, which has little significance. Properly managed grazing animals are an important part of the solution to feeding the world in the future.

I may once have harbored a notion that by following a vegetarian diet I was choosing a path that ensured nothing would have to die for my meals. But no longer. The more I've become familiar with agriculture, the more that seems a gross oversimplification. As I've studied the globe's hundreds of millions of years of change, it's readily apparent that the earth was never assaulted on such a broad scale until the onset of crop agriculture, an effect that was greatly multiplied with the advent of mechanization. Pastoral animal keeping, however, mimics the functions of wild herds that covered the earth for millions of years. The impact of animal herds is a *familiar* disturbance to the earth, one that plants and animals can tolerate and actually *need*. Vegetation will be pruned and stepped on, as it has always been, allowing for a diversity of plants, and for perennialism. Compared with other ways of producing food, the keeping of grazing livestock, when done appropriately, is the most environmentally benign. The best lives for domesticated animals are on grass, and grass provides the most opportunities for wild animals of all shapes and sizes. Raising cattle on grass thus provides habitat for both the domesticated and the wild.

My primary mission this past decade has been helping, however possible, to build a more environmentally sound and humane food system. There is such a terribly long way to go. I don't urge people to eat meat. But for those who do, I encourage them to seek meat that is well raised. At the same time, I don't consider abandoning meat an effective strategy for positively affecting the food system. Instead, I believe the most important thing a consumer can do to change the way meat is produced is to *buy meat* from well-raised animals. In other words, to directly support farmers and ranchers who are doing things the right way, nearly all of whom are operating on extremely thin margins. In a nation where agribusiness, food, and pharmaceutical companies hold the political cards, I hold out little hope for any major policy reform. Consumers, however, have enormous power to make positive change in the food system.

I see the raising of cattle and other grazing animals on grass, which is inherently resistant to industrialization, as an essential part of a more locally and regionally based, more environmentally sustainable food system. Necessarily, livestock grazing is broadly dispersed, outdoors, and reliant on natural systems. It requires local people, out on the land, with knowledge of climate and ecosystem functions. This is why traditional livestock tenders still exist the world over while croplands have been largely co-opted by multinational agribusiness corporations growing commodities for the international market.

In the United States, I think this is what alarms agribusiness the most about the rising interest in totally grass-fed beef. The huge corporations have gained near-absolute control over the beef markets by buying, shutting down, and consolidating feedlots and slaughterhouses. Although American cattle ranchers remain fiercely independent, they are price takers, not price makers, because they control none of the infrastructure that turns cattle into meat and gets it to the end user. Yet the industry does not have, and probably can never have, similar control over cattle raised entirely on grass. This is the province of a growing number of independent farmers and ranchers scattered around the United States who mostly raise their own animals from birth and sell beef directly to consumers.

On another level, I support cattle ranching and farming because I believe humans' age-old association with farm animals provides important intangible benefits to humanity. We are better for living alongside them. Those of us who have the pleasure of being around them every day likely benefit the most. We are taught in stark relief the lessons of nature—the inevitability of illness, injury, and death; the cycles of birth, growth, aging, and decline. We are constantly reminded of the fragility of life, of what it takes to be a good parent, of bravery, patience, loyalty. If we are paying attention, we are learning from them, constantly. Our own impermanence is clearer to us. They also bring unquantifiable yet vast pleasure to many of the people who see them, as evidenced by the frequency with which people slow down their cars or even stop and get out just to observe our animals. I have been told again and again by people in our community how much they enjoy seeing and being near them. "I love walking among your cows," a neighbor told me once. "They make me feel so calm."

Humans are strongly drawn to meat and other foods derived from animals. Research demonstrates that as people gain affluence, to a point, they increase the meat and dairy in their diet. This is happening today in Asia, where demand for meat is growing the fastest. Research also shows that about 75 percent of people who give up meat return to it eventually. I've never seen research on this point, but I strongly suspect the percentage would be even higher for people who've attempted a strictly vegan diet. I've met dozens of former vegans or vegetarians in my lifetime. Whether for reasons of culture, health, or convenience, the overwhelming majority of the world's people eat foods from animals. Clearly, humans will be raising animals for food for the foreseeable future.

Our food system must regenerate itself as nature does, and treat animals as our partners, not as inanimate production units. We must restore the broken covenant we have with farm animals. Ridding the world of factory farms should be a priority. Industrial operations must be replaced with farming systems that recycle nutrients and continually feed the earth's living blanket, the soils.

At the heart of this regenerative food system will be cattle and other grazing animals. We will manage their movements, help protect them from predators, make sure they have water and good forage, and attend to their needs. In return, they will help us by keeping our grasslands vibrant—covered in vegetation and teeming with carbon and life belowground. Our cattle will provide us milk and meat, healthful, nutrient-dense foods. We will appreciate and value them for all that they are and all they do.

ACKNOWLEDGMENTS

This book has been years in the making, and could never have happened without the assistance of many colleagues, friends, and family members along the way. Several people have, over the past decade, generously and patiently shared with me their knowledge of cattle husbandry. To each of you, I want to express my enormous admiration and respect for the honorable way you practice your important profession. In particular, I want to thank: Annie Van Peer, Rob and Michelle Stokes, Ken Bentz Sr., Don McNab, and of course Bill Niman.

Several people kindly took time from their busy schedules to read parts of this manuscript or helped me track down pieces of information. I am indebted to each of you, and am truly grateful for the many valuable suggestions each of you made: Bill Niman, Christine Hahn, Gareth Fisher, and Adam Danforth.

I want to thank my agent, Jennifer Unter, who has unfailingly represented me in the most attentive and professional way; my editor, Ben Watson, who believed in this book from the start, and was responsive, helpful, and professional throughout; and copy editor Laura Jorstad for her careful and thorough review.

Most of all I want to thank my wonderful family, Bill, Miles, and Nicholas, each of whom had to get along without me, or with me being distracted, at various times as I toiled away on this book. Bill often did double duty so I would have time to work on the book. Miles often had to go without the cherished bedtime stories I've read him every night since he was three months old. And Nicholas, more often than not, was nursed in my lap as I researched or edited my writing. I could never have accomplished this task without the patience, love, and support I feel from each of you every day. I am extremely grateful and I love each of you with my whole heart.

NOTES

Introduction

1. Snapp, R. R. *Beef Cattle: Their Feeding and Management in the Corn Belt States*, 3rd ed. (Wiley & Sons, 1939), 16.
2. See: Hahn Niman, N. "Beef, the Most (Unfairly) Maligned of Meats." Chapter 7 in *Righteous Porkchop: Finding a Life and Good Food Beyond Factory Farms* (HarperCollins, 2009).
3. Ibid., chapter 8, "The (Un)-Sacred Milk Cow."
4. The United States is actually fourth (after India, Brazil, and China) in the number of cattle, but it produces the most beef. *BEEF Magazine* (August 13, 2013). http://beefmagazine.com/cattle-industry-structure/industry -glance-beef-cow-inventory-over-time.

Chapter 1: The Climate Change Case Against Cattle

1. The headline for the press release of *Livestock's Long Shadow* read: "Rearing Cattle Produces More Greenhouse Gases than Driving Cars." www.un.org/apps /news/story.asp?newsID=20772&CR1=warning#.UpjwIo2f_eI.
2. Hahn Niman, N. "The Carnivore's Dilemma." *New York Times* (October 30, 2009). www.nytimes.com/2009/10/31/opinion/31niman. html?pagewanted=all.
3. "Livestock Manure Management," a report of the U.S. Environmental Protection Agency. www.epa.gov/methane/reports/05-manure.pdf.
4. Hahn Niman, N. "Animals Are Essential to Sustainable Food." *Earth Island Journal* (Spring 2010). www.earthisland.org/journal/index.php /eij/article/rancher; cattle numbers are from the 2012 U.S. Department of Agriculture Census of Agriculture (2014). www.agcensus.usda.gov /Publications/2012/#full_report.
5. "Tackling Climate Change Through Livestock," a report of the United Nations Food and Agriculture Organization (September 2013). www.fao.org/docrep /018/i3437e/i3437e.pdf.
6. Leng, R. A. *Quantitative Ruminant Nutrition: A Green Science* (1993). www.ciesin .columbia.edu/docs/004-180/004-180.html.
7. Ibid.
8. Re: dung beetles, see: www.sciencedaily.com/releases 2013/08/130822105031. htm; re: feeding and genetics for lower methane, see: www.sciencedaily.com/ releases/2014/01/140110131013.htm; re: weights in the rumen, see: "Okine et al. have demonstrated a 29% reduction in methane production when weights were added to the rumen to stimulate contraction of the rumen wall in order to decrease residence time of the feed in the digestive tract." Okine, E. K., Mathison, G. W., and Hardin, R. T. "Effects of Changes in Frequency of Reticular Contractions on Fluid and Particulate Passage Rates in Cattle." *Canadian Journal of Animal Science* 67 (1989), 3388; re: minerals, see: "Adequate mineral supplementation is another avenue by which the cattle industry may realize

a net reduction in enteric methane emissions." Ominski, K., and Wittenberg, K. "Strategies for Reducing Enteric Methane in Forage-Based Beef Production Systems," paper presented at Science of Changing Climates conference, Edmonton, Alberta (July 2004), 13. www1.foragebeef.ca/$foragebeef/frgebeef .nsf/all/ccf758/$FILE/CcStrategiesforreducingemmissionsOminski.pdf.

9. Schwartz, J. *Cows Save the Planet* (Chelsea Green Publishing, 2013), 27.

10. Website of the U.S. Environmental Protection Agency. http://epa.gov/climate change/ghgemissions/gases/n2o.html.

11. "Fertilizer Use Responsible for Increase in Nitrous Oxide in Atmosphere," news release of the UC Berkeley News Center (April 2, 2012). https://newscenter.berkeley.edu/2012/04/02/ fertilizer-use-responsible-for-increase-in-nitrous-oxide-in-atmosphere.

12. *Drawing Down N₂O to Protect Climate and the Ozone Layer*, a United Nations Environment Program report (November 2013), 21. www.unep.org/pdf/ UNEPN2Oreport.pdf.

13. Fairlie, S. *Meat: A Benign Extravagance* (Chelsea Green Publishing, 2011), 159.

14. Website of the U.S. Environmental Protection Agency. http://epa.gov/climate change/ghgemissions/sources/agriculture.html.

15. Fairlie, 161.

16. Ibid.

17. *The Emissions Gap Report, 2013*, a United Nations Environment Program report (November 2013), xvi. www.unep.org/publications/ebooks/emissionsgap report2013/portals/50188/EmissionsGapReport%202013_high-res.pdf.

18. "Livestock and Climate Change." *World Watch* (November–December 2009). www.worldwatch.org/files/pdf/Livestock%20and%20Climate%20Change.pdf.

19. From an interview with Robert Goodland. http://juliansstory.wordpress. com/2011/10/27/meeting-robert-goodland.

20. www.spectrumcommodities.com/education/commodity/statistics/soybeans .html; and www.spectrumcommodities.com/education/commodity/statistics /soybeantable.html.

21. "How Much U.S. Meat Comes from Foreign Sources," a report of the Economic Research Service of the U.S. Department of Agriculture (September 20, 2012). www.ers.usda.gov/amber-waves/2012-september/how-much-us-meat.aspx# .UrDRko0hYVl.

22. Hahn Niman, *Righteous Porkchop*, 136–40.

23. Biswell, H. *Prescribed Burning in California Wildlands Vegetation Management* (University of California Press, 1989).

24. Ibid., 48.

25. Ibid., 49–50.

26. Mann, C. *1491* (Vintage, 2006), 285–86.

27. Montgomery, D. *Dirt: The Erosion of Civilizations* (University of California Press, 2012), 29.

28. See: ibid.

29. *U.S. Forest Facts and Historical Trends*, a report of the Forest Service of the U.S. Department of Agriculture (September 2001). www.fia.fs.fed.us/library/ briefings-summaries-overviews/docs/ForestFactsMetric.pdf.

30. "Deforestation." *Encyclopedia Britannica*. www.britannica.com/EBchecked/ topic/155854/deforestation/306437/Effects.

31. Palmer, L. "In the Pastures of Colombia, Cows, Crops and Timber Coexist." *Yale Environment 360* (March 13, 2014). http://e360.yale.edu/feature/ in_the_pastures_of_colombia_cows_crops_and_timber_coexist/2746.

32. See, e.g.: Spiegal, S., Huntsinger, L., Hopkinson, P., Bartolome, J., and Sheri, S. "Overview of California Range Ecosystems." In *Ecosystems of California*, edited by H. Mooney and E. Zavaleta (University of California Press, in press, due August 2015).

33. Azeez, G. "Soil Carbon and Organic Farming: Summary of Findings," a report of the UK Soil Association (2009). www.soilassociation.org/LinkClick. aspx?fileticket=BVTfaXnaQYc%3D&.

34. Mitloehner, F. M. "Clearing the Air on 'Livestock and Climate Change'" (2010). www.ansci.cornell.edu/cnconf/2010proceedings/CNC2010.5.Mitloehner.pdf.

35. Schwartz, 12.

36. "Tackling Climate Change," FAO, 41.

37. Ibid., xiii (emphasis added).

38. See, e.g.: Gattinger, A., et al. "Enhanced Top Soil Carbon Stocks Under Organic Farming." *Proceedings of the National Academy of Sciences* (October 2012). www. pnas.org/content/109/44/18226.full.

39. Cited in Azeez: "An estimated 89% of the global potential for agricultural greenhouse gas mitigation would be through carbon sequestration." Smith, P., et al. "Greenhouse Gas Mitigation in Agriculture." *Philosophical Transactions of the Royal Society of London, Series B Biological Sciences* (2008).

40. Azeez.

41. Montgomery, 23.

42. Ibid., 3.

43. Schwartz, J. "Soil as Carbon Storehouse: New Weapon in Climate Fight?" *Yale Environment 360* (March 4, 2014). http://e360.yale.edu/feature/soil_as _carbon_storehouse_new_weapon_in_climate_fight/2744.

44. Azeez.

45. Ibid. See also: "[A] feedback mechanism may exist wherein increased global warming intensifies rainfall, which, in turn, increases erosion and continues the cycle." Pimentel, D., and Pimentel, M. *Food, Energy, and Society*, 3rd ed. (CRC Press, 2008), 212, citing Lal, 2002.

46. Ibid.

47. "Glomalin: Hiding Place for a Third of the World's Stored Soil Carbon." *Agricultural Research*, a publication of the U.S. Department of Agriculture (September 2002). www.ars.usda.gov/is/ar/archive/sep02/soil0902.htm; "Glomalin: What Is It . . . and What Does It Do?" *Agricultural Research* (July 2008). www.ars.usda.gov/is/AR/archive/jul08/glomalin0708.pdf.

48. See, e.g.: Fonte, S., et al. "Earthworm Populations in Relation to Soil Organic Matter Dynamics and Management in California Tomato Cropping Systems." *Applied Soil Ecology* (2009). http://ucanr.org/sites/ct/files/44375.pdf.

49. Azeez.

50. USDA *Agricultural Research* articles.

51. Personal communication with author, April 22, 2014.

52. Azeez.

53. Ibid.

54. Manske, L. "Effects of Grazing Management Treatments on Rangeland Vegetation." North Dakota State University, website of the Dickinson Research Extension Center (2004). www.ag.ndsu.nodak.edu/dickinsonresearch/2003 /range03c.htm.

55. Harnett, D. Paper presented to the 1999 Society for Range Management Meeting (February 1999). Website of the Ecological Society of America, www.esa.org /science_resources/publications/purePrairieLeague.php.

56. Hahn Niman, "Animals Are Essential."

57. Lal, R., and Stewart, B. A. *Food Security and Soil Quality* (CRC Press, 2010), 255. www.academia.edu/5208901/ Food_Security_and_Soil_Quality_By_Rattan_Lal_B_A._Stewart.
58. Ibid., 263.
59. Ibid., 251.
60. Guo, L. B., and Gifford, R. M. "Soil Carbon Stocks and Land Use Change: A Meta Analysis." *Global Change Biology* (November 23, 2002). http://onlinelibrary. wiley.com/doi/10.1046/j.1354-1013.2002.00486.x/abstract.
61. Azeez.
62. Schwartz, *Cows Save the Planet*, 15; and Schwartz, "Soil as Carbon Storehouse."
63. Schwartz, *Cows Save the Planet*, 15.
64. "Creating Topsoil by Dr. Christine Jones" (March 24, 2006). http://creatingnew soil.blogspot.com.
65. Savory, A. "Reversing Global Warming While Meeting Human Needs: An Urgently Needed Land-Based Option," speech given at Tufts University (January 25, 2013).
66. Ibid.
67. Ibid.
68. "The Savory Institute: Healing the World's Grasslands, Rangelands, and Savannas." Allan Savory interview with *World Watch* (April 15, 2011). http://blogs.worldwatch.org/nourishingtheplanet/tag/ international-society-for-range-management.
69. Savory, "Reversing Global Warming."
70. Ibid.
71. Marty, J. "Effects of Cattle Grazing on Diversity in Ephemeral Wetlands." *Conservation Biology* 19, no. 5 (October 2005), *citing* White, 1979; Sousa, 1984; Hobbs and Huenneke, 1992. www.elkhornsloughctp.org/uploads/ files/1401307451Marty.%202005.%20Effects%20of%20cattle%20grazing%20 on%20diversity%20in%20ephemeral%20wetlands..pdf.
72. Savory, *World Watch* interview.
73. Savory, A. *Holistic Resource Management* (Island Press, 1988), 38.
74. Ibid., 35, 38.
75. Ibid., 39–40.
76. Ibid., 37.
77. Ibid.
78. Ibid., 38.
79. Ibid., 37.
80. Savory, A. Talk given at the Restoring Working Landscapes, Producing Sustainable Meat conference, TomKat Ranch, Pescadero, California (December 3, 2013).
81. Savory, "Reversing Global Warming."
82. Schwartz, J. "Soil as Carbon Storehouse."
83. Itzkan, S. "Hut with a View," Seth Itzkan Reports from the Africa Centre for Holistic Management in Zimbabwe. http://hutwithaview.com.
84. Savory, *Holistic Resource Management*, 9.
85. Savory, TomKat Ranch.
86. Savory, *Holistic Resource Management*, 34–35.
87. Savory, "Reversing Global Warming."
88. Savory, *World Watch* interview.
89. University of Maryland press release (March 10, 2014). www.umdrightnow.umd .edu/news/animals-key-biodiversity-over-fertilized-prairies.

90. Savory, TomKat Ranch.
91. Savory, "Reversing Global Warming."
92. Savory, *World Watch* interview.
93. Savory, TomKat Ranch.
94. Savory, *World Watch* interview.
95. Savory, TomKat Ranch.
96. Ibid.
97. Savory, "Reversing Global Warming."
98. Teague et al. "Grazing Management Impacts on Vegetation, Soil Biota and Soil Chemical, Physical and Hydrological Properties in Tall Grass Prairie." *Agriculture, Ecosystems and Environment* (2011). www.sciencedirect.com/ science/article/pii/S0167880911000934.
99. Weber et al. *Journal of Arid Environments* 141 (May 2011), 310. www.science direct.com/science/article/pii/S0140196310003460.
100. Itzkan, S. "Upside (Drawdown): The Potential of Restorative Grazing to Mitigate Global Warming by Increasing Carbon Capture on Grasslands," a report of PlanetTech Associates (April 2014). www.planet-tech.com/upsidedrawdown# sthash.mKkNI9fj.dpuf.
101. Coleman, E. "Debunking the Meat/Climate Change Myth" (August 7, 2009). http://grist.org/article/2009-08-07-debunking-meat-climate-change-myth.
102. "Tackling Climate Change," FAO.
103. Ibid., 23.
104. Ibid., 17.
105. Lal and Stewart, 260.
106. Ibid., 184.
107. Tuhus-Dubrow, R. "How to Solve Climate Change with Cows (Maybe). Could Better Grazing Patterns Be the Answer? A Sweeping New Theory Divides the Environmental World." *Boston Globe* (May 4, 2014). www.bostonglobe.com/ ideas/2014/05/03/how-solve-climate-change-with-cows-maybe/j3c4uoHv4 iJqWjHXezlonN/story.html.
108. "Tackling Climate Change," FAO, 17.
109. Rosenthal, E. "To Cut Global Warming, Swedes Study Their Plates." *New York Times* (October 22, 2009). www.nytimes.com/2009/10/23/world/europe/23 degrees.html?pagewanted=1&_r=3&.
110. Hendrickson, J. *Energy Use in the U.S. Food System: A Summary of Existing Research and Analysis*, a report of The Center for Integrated Agricultural Systems, University of Wisconsin–Madison (July 2008). www.cias.wisc.edu /wp-content/uploads/2008/07/energyuse.pdf.
111. *Organic Works: Providing More Jobs Through Organic Farming and Local Food Supply*, a report of the UK Soil Association (Bristol House, 2006).
112. Note that "Tackling Climate Change," FAO, 17, states: "Emissions associated with energy consumption (directly or indirectly related to fossil fuel) are mostly related to feed production, and fertilizer manufacturing, in particular. When added up along the chains, energy use contributes about 20 percent of total sector emissions."
113. Murray, S. "The Deep Fried Truth." *New York Times* (December 14, 2007). www.nytimes.com/2007/12/14/opinion/14murray. html?module=Search&mabReward=relbias%3Ar.
114. Hahn Niman, N., and Niman, B. "The Cost of Wasted Food." *The Atlantic* (December 2, 2009). www.theatlantic.com/health/archive/2009/12/the-cost -of-wasted-food/31089.

115. "World War II Rationing." *United States History*. www.u-s-history.com/pages/h1674.html.

Chapter 2: All Food Is Grass

1. For the 40 percent figure, see World Resources Institute, www.wri.org/publication/content/8269. For the 70 percent figure, see: "Are Grasslands Under Threat?," an analysis of the United Nations Food and Agriculture Organization. www.fao.org/uploads/media/grass_stats_1.pdf.
2. Stromberg et al., eds. *California Grasslands: Ecology and Management* (University of California Press, 2007), 7.
3. Pimentel and Pimentel, 363.
4. Montgomery, x.
5. Ibid., xiv.
6. Ibid., 16.
7. Ibid., 15.
8. Rogers, J., and Feiss, P. G. *People and the Earth: Basic Issues in the Sustainability of Resources and Environment* (Cambridge University Press, 1998), 63. See also: "About one-third of the topsoil from U.S. agricultural land already has been lost," Pimentel and Pimentel, 41.
9. Curry, J. P. *Grassland Invertebrates: Ecology, Influence on Soil Fertility and Effects on Plant Growth* (Springer, 1993).
10. Pimentel and Pimentel, 190.
11. Montgomery, 17. See also: "Along with plants and animals, microbes are a vital component of the soil and constitute a large percentage of the soil biomass. One square meter of soil may support about 200,000 arthropods and enchytraeids, plus billions of microbes (Wood, 1989; Lee and Foster, 1991) . . . In addition, soil bacteria and fungi add 4000–5000 species and in this way contribute significantly to the biodiversity. . . ." Pimentel and Pimentel, 209.
12. Troeh, F., et al. *Soil and Water Conservation*, 3rd ed. (Prentice-Hall, 1999), 332. See also: "Vegetative cover is the principal way to protect soil and water resources." Pimentel and Pimentel, 31.
13. Ibid.
14. Lal and Stewart, 152. See also: "[E]roded soil absorbs 87% less water by infiltration than uneroded soils." Pimentel and Pimentel, 190, citing Guenette, 2001.
15. Pimentel and Pimentel, 203.
16. Berry, W. *New Roots of Agriculture* (University of Nebraska Press, 1980), xii.
17. Pimentel and Pimentel, 31.
18. Berry, xiii.
19. Pimentel and Pimentel, 205.
20. Ibid.
21. Ibid., 206.
22. Boody, G., et al. "Multifunctional Agriculture in the United States." *Bioscience* 55 (2005), 32.
23. Lal and Stewart, 153 (internal citation omitted).
24. See: "Fundamentals of Cation Exchange Capacity." https://www.extension.purdue.edu/extmedia/ay/ay-238.html.
25. Lal and Stewart, 252.
26. Genesis 3:19.
27. Psalms 104.
28. Deuteronomy 11:15.

29. Ingalls, J. J. "In Praise of Bluegrass" (1872). http://grassbydesign.com/pdf/douglas.pdf.
30. *California Grasslands*, 37.
31. "Last Time Carbon Dioxide Levels Were This High: 15 Million Years Ago." *Science Daily* (October 9, 2009). www.sciencedaily.com/releases/2009/10/091008152242.htm.
32. *California Grasslands*, 37.
33. Ibid., 50–52.
34. Ibid., 49–52.
35. Kinver, M. "Ecologists Learn Lessons from the Ghosts of Megafauna." *BBC News* (March 24, 2014). www.bbc.com/news/science-environment-26718199.
36. "Big Game Could Roam U.S. Plains." *BBC News* (August 18, 2005). http://news.bbc.co.uk/2/hi/science/nature/4160560.stm.
37. Daley, S. "From Untended Farmland, Reserve Tries to Recreate Wilderness from Long Ago." *New York Times* (June 13, 2014). www.nytimes.com/2014/06/14/world/europe/from-untended-farmland-reserve-tries-to-recreate-wilderness-from-long-ago.html?emc=edit_th_20140614&nl=todaysheadlines&nlid=28644965&_r=0.
38. Ibid.
39. *California Grasslands*, 2.
40. See: ibid., 57, 58, 59, 64, 65, 75, 175, 209, 219, 256.
41. Hahn Niman, *Righteous Porkchop*, 136, *citing* re: cattle zoological classification, Hickman, C. *Integrated Principles of Zoology*, 7th ed. (Times Mirror/Mosby College Publishing, 1994), 662; re: domestication, Davis, P., and Dent, A. *Animals That Changed the World: The Story of Domestication of Wild Animals* (Crowell-Collier Press, 1968), 66, and Beja-Pereria, A., et al. "The Origin of European Cattle: Evidence from Modern and Ancient DNA." *Proceedings of the National Academy of Sciences* (May 11, 2006).
42. Snapp, 5, *citing Yearbook of Agriculture 1921*, U.S. Department of Agriculture (U.S. Government Printing Office, 1921), 232.
43. Snapp, 16.
44. Montgomery, 18.
45. Hughes, H. D. *Forages* (Iowa State College Press, 1951), 22.
46. Anderson, C. P. *Grass: Yearbook of Agriculture 1948*, U.S. Department of Agriculture (U.S. Government Printing Office, 1948), v.
47. Hughes, 8.
48. Montgomery, 173; Pimentel and Pimentel, 41.
49. Pimentel, D., and Kounang, N. "Ecology of Soil Erosion in Ecosystems." *Ecosystems* 1 (1998), 416.
50. "Losing Ground," a report of the Environmental Working Group (April 2011). http://static.ewg.org/reports/2010/losingground/pdf/losingground_report.pdf.
51. Barber, D. "What Farm-to-Table Got Wrong." *New York Times* (May 17, 2014).
52. See, e.g.: Hahn Niman, N., "Biofuels: Bad News for Animals." *AWI Quarterly* (Summer 2008). https://awionline.org/content-types-orchid-legacy/awi-quarterly/biofuels-bad-news-animals.
53. Zuckerman, J. "Plowed Under." *American Prospect* (May 2014). http://prospect.org/article/plowed-under.
54. Kreiger, L. "California Drought: San Joaquin Valley Sinking as Farmers Race to Tap Aquifer." *San Jose Mercury News* (March 29, 2014). www.mercurynews.com/drought/ci_25447586/california-drought-san-joaquin-valley-sinking-farmers-race.

55. See: Thicke, F. *A New Vision for Iowa Food and Agriculture* (Mulberry Knoll Books, 2010).
56. Lal and Stewart, 130.

Chapter 3: Water

1. Dell'Amore, C. "Biggest Dead Zone Ever Forecast in Gulf of Mexico: Oxygen-Deprived Area May Be Size of New Jersey, Scientists Say." *National Geographic News* (June 24, 2013). http://news.nationalgeographic.com/news/2013/06/130621-dead-zone-biggest-gulf-of-mexico-science-environment.
2. See, e.g.: Hahn Niman, *Righteous Porkchop*, 49–51.
3. See: Charles, T., and Stuart, H. *Commercial Poultry Farming* (Interstate Printing, 1936), v.
4. Pimentel and Pimentel, 232.
5. Ibid., 210.
6. Jackson, W. *New Roots for Agriculture* (University of Nebraska Press, 1980), 20.
7. See: Burkholder, J. "Impacts of Waste from Concentrated Animal Feeding Operations on Water Quality." *Environmental Health Perspectives* (February 2007). www.ncbi.nlm.nih.gov/pmc/articles/PMC1817674.
8. "Concentrated Animal Feeding Operations: Health Risks from Water Pollution," a publication of the Institute for Agriculture and Trade Policy (IATP). www.iatp.org/files/421_2_37390.pdf.
9. Today the vast majority (over 90 percent) of American livestock and poultry feed is corn. You often see the figure that 70 percent of U.S. grain is fed to our livestock and poultry, but this figure is out of date. About 40 percent of corn grown in the United States is now being used for ethanol. A large portion is also exported. See: Foley, J. "It's Time to Rethink America's Corn System." *Scientific American* (March 5, 2013). www.scientificamerican.com/article/time-to-rethink-corn; and Hoffman, L., et al. "Feed Grains Backgrounder," a report of the Economic Research Service of the U.S. Department of Agriculture (March 2007). www.ers.usda.gov/media/197317/fds07c01_1_.pdf.
10. Thompson, S. "Running on Empty?: 'Great ethanol debate' waged at NCGA forum." *Rural Cooperatives Magazine*, a publication of the U.S. Department of Agriculture (Sept/Oct 2005), http://www.rurdev.usda.gov/rbs/pub/sep05/running.htm.
11. Pimentel and Pimentel, 317.
12. "Fertilizer Use and Price," a report of the U.S. Department of Agriculture, Economic Research Service (July 12, 2013). www.ers.usda.gov/data-products/fertilizer-use-and-price.aspx#.U3AfAy_cOXw.
13. "Putting Dairy Cows Out to Pasture: An Environmental Plus," a report of the U.S. Department of Agriculture, Agricultural Research Service (April 29, 2011). www.ars.usda.gov/is/ar/2011/may11/cows0511.htm.
14. Hendrickson; "Agriculture's Supply and Demand for Energy and Energy Products," a report by the U.S. Department of Agriculture, Economic Research Service (May 2013). www.ers.usda.gov/media/1104145/eib112.pdf.
15. Ibid.
16. Ibid.
17. Pimentel and Pimentel, 153.
18. Ibid., 161.
19. "Pesticides in the Nation's Streams and Groundwater," a fact sheet of the U.S. Geological Survey (March 2006). http://pubs.usgs.gov/fs/2006/3028 (emphasis added).
20. Ibid.

21. Ibid.
22. Nichols, K. "The Role of Soil Biology in Improving Soil Quality," a webinar of National Resources Conservation Service /U.S. Department of Agriculture (September 13, 2012). https://www.youtube.com/watch?v=eGxjcxVMbsg.
23. "Concentrated Animal Feeding Operations," IATP.
24. See, e.g., the Pew Charitable Trusts' Campaign on Human Health and Industrial Farming. www.pewtrusts.org/en/projects/campaign-on-human-health-and-industrial-farming.
25. Dean, C. "Water Pollution, Drugs and Household Productions in the Water Supply, Does It Matter?" *New York Times* (April 3, 2007). www.nytimes.com/2007/04/03/science/earth/03water.html?pagewanted=all&_r=0.
26. See: Hahn Niman, *Righteous Porkchop*, chapter 8.
27. Winsten, J. R., et al. "Differentiated Dairy Grazing Intensity in the Northeast." *Journal of Dairy Science* 83 (2000), 836.
28. Ibid.
29. Halverson, M. *Farm Animal Health and Well-Being*, a report prepared for the Minnesota Planning Agency, Environmental Quality Board (April 23, 2001), 120.
30. Marks, R. *Cesspools of Shame*, a report by the Natural Resources Defense Council (July 2001), 11. www.nrdc.org/water/pollution/cesspools/cesspools.pdf.
31. Whitlock, R. *A Short History of Farming in Britain* (John Baker Ltd., 1965).
32. Snapp, 16.
33. Data from the 2012 U.S. Department of Agriculture Census of Agriculture (May 2014). www.agcensus.usda.gov/Publications/2012.
34. *Ecosystems of California*, 37 (internal citation omitted).
35. "Sustainable Farming Systems: Demonstrating Environmental and Economic Performance," a study of the University of Minnesota et al. (June 2001).
36. From the vegan advocacy website vegsource.com. "How Much Water to Make One Pound of Beef" (accessed March 2013).
37. Smithers, R. "Food Products Should Carry 'Water Footprint' Information, Says Report." *Guardian* (July 20, 2009). www.guardian.co.uk/environment/2009/jul/20/food-water-footprint.
38. From the website vegsource.com, *citing* Pimentel, D., Westra, L., and Noss, R. F. (eds.). *Ecological Integrity: Integrating Environment, Conservation, and Health* (Island Press, 2001).
39. Beckett and Oltjen. "Examination of the Water Requirement for Beef Production in the United States." *Journal of Animal Science* 71 (1993), 818.
40. Keith, L. *The Vegetarian Myth* (PM Press, 2009), 102. See also: "[A]nimal protein has about 1.4 times the biological value as food compared with grain protein." Pimentel and Pimentel, 70.
41. Ibid., 103.
42. Boody, 32.
43. "Managing Runoff and Erosion on Croplands and Pastures," a publication of the University of Georgia Cooperative Extension (March 2009). www.caes.uga.edu/applications/publications/files/pdf/B%201152-15_2.PDF.

Chapter 4: Biodiversity

1. Steinfeld, H., et al. *Livestock's Long Shadow: Environmental Issues and Options*, a report of the United Nations Food and Agriculture Organization (2007), 254.
2. Schwartz, *Cows Save the Planet*, 193.
3. Krausman, P. "An Assessment of Rangeland Activities on Wildlife Populations and Habitats," a publication of the National Resource Conservation Service (2011),

257. www.nrcs.usda.gov/Internet/FSE_DOCUMENTS/stelprdb1045801.pdf (internal citations omitted).

4. Ibid., 259.

5. Ibid., *citing* Licht, 1997; Higgins et al., 2002.

6. Marty, *citing* Noy-Meir et al., 1989; Harrison, 1999; Hayes and Holl, 2003. http://vernalpools.net/documents/Marty%20Cons%20Bio.pdf.

7. Ibid., *citing* McNaughton et al., 1989; Milchunas and Lauenroth, 1993; Perevolotsky and Seligman, 1998.

8. Ibid., *citing* Collins et al., 1998; Harrison, 1999; Maestas et al., 2003.

9. "Animals Key to Biodiversity of Over-Fertilized Prairies" (March 10, 2014). www .umdrightnow.umd.edu/news/animals-key-biodiversity-over-fertilized-prairies; see also: "How Grazing Helps Plant Diversity" (2014). www.sciencedaily.com /releases/2014/03/140309150536.htm?utm_source=hootsuite&utm_campaign =hootsuite#.U2CJmC0cly0.facebook.

10. Hart, R. "Plant Biodiversity on Shortgrass Steppe After 55 Years of Zero, Light, Moderate, or Heavy Cattle Grazing." *Plant Ecology* (July 2001). http://link .springer.com/article/10.1023/A:1013273400543.

11. "Vegetation Change After 65 Years of Grazing and Grazing Exclusion." *Journal of Rangeland Management* (November 2004). www.cabnr.unr.edu/news/story .aspx?StoryID=295.

12. Ibid.

13. *Ecosystems of California*, 23.

14. "Vernal Pools: Liquid Sapphires of the Chaparral." Website of The Chaparral Institute. www.californiachaparral.org/vernalpools.html.

15. Marty.

16. The Nature Conservancy newsletter (Fall 2009). www.nature.org/ourinitiatives /regions/northamerica/unitedstates/california/fall-2009-newsletter-1-1.pdf.

17. Krausman, 262.

18. Marty (internal citations omitted).

19. "Grazing CRP Land Improves Feed, Habitat," a case study of the Leopold Center for Sustainable Agriculture. www.leopold.iastate.edu/sites/default/files/Grazing CRPcasestudy.pdf.

20. Hahn Niman, "Biofuels: Bad News for Animals."

21. Krausman, 262, *citing* Kohler and Rauer, 1991.

22. *Ecosystems of California*, 30–31.

23. U.S. Fish and Wildlife Service data. www.fws.gov/pollinators.

24. Data taken from a report by biologists at UC Berkeley. See my article on the study: Hahn Niman, N. "A Way to Save America's Bees: Buy Free Range Beef." *The Atlantic* (July 14, 2011). www.theatlantic.com/health /archive/2011/07/a-way-to-save-americas-bees-buy-free-range-beef/241935.

25. Wines, M. "Soaring Bee Deaths in 2012 Sound Alarm on Malady." *New York Times* (March 28, 2013). www.nytimes.com/2013/03/29/science/earth /soaring-bee-deaths-in-2012-sound-alarm-on-malady.html.

26. Hahn Niman, "A Way to Save America's Bees."

27. Ibid.

28. Krausman, 279.

29. Mendenhall, C., et al. "Predicting Biodiversity Change and Averting Collapse in Agricultural Landscapes." *Nature* (May 8, 2014). www.nature.com/nature /journal/vaop/ncurrent/full/nature13139.html.

Chapter 5: Overgrazing

1. Montgomery, 236.
2. Ibid., xii.
3. Lal, R. "Potential of Desertification Control to Sequester Carbon and Mitigate the Greenhouse Effect." *Climatic Change* 51, no. 1 (October 2001), 35–72.
4. Rosen, S. "Desertification and Pastoralism: A Historical Review of Pastoral Nomadism in the Negev Region." Encyclopedia of Life Support Systems.
5. Montgomery, xx.
6. Ibid., xii.
7. Ibid., 21.
8. Ibid., 4.
9. Jackson, 2.
10. Ibid.
11. Ibid., 12.
12. Montgomery, 5.
13. Ibid., 23.
14. Ibid., 23–24.
15. Jackson, 12 13.
16. Starrs, P. *Let the Cowboy Ride* (Johns Hopkins University Press, 1998), 21.
17. Ibid.
18. "Fact Sheet on the BLM's Management of Livestock Grazing," a document of the U.S. Bureau of Land Management. www.blm.gov/wo/st/en/prog/grazing.html.
19. Decision Notice and Finding of No Significant Environmental Impact. Forest Service of the U.S. Department of Agriculture (September 2010), 59. www.fs.usda.gov/Internet/FSE_DOCUMENTS/stelprdb5200869.pdf (internal citations omitted).
20. Mearns, R. "Livestock and Environment: Potential for Complementarity." www.fao.org/docrep/w5256t/w5256t02.htm.
21. Weber, K., and Horst, S. "Desertification and Livestock Grazing: The Roles of Sedentarization, Mobility and Rest." *Pastoralism* (October 2011), *citing* Seligman and Perevolotsky, 1994; Olaizola et al., 1999; Cummins, 2009. www.pastoralism journal.com/content/1/1/19#.
22. See: Lal and Stewart.
23. Montgomery, 208.
24. Sorensen, L. "Colorado Cattle Stomp Shows the Benefit of Healing Hooves." *BEEF Magazine* (May 7, 2014). http://m.beefmagazine.com/pasture-health/colorado-cattle-stomp-shows-benefit-healing-hooves?utm_content=buffera01bd&utm_medium=social&utm_source=twitter.com&utm_campaign=buffer.

Chapter 6: People

1. Louv, R. *Last Child in the Woods: Saving Our Children from Nature Deficit Disorder* (Algonquin Books, 2006).
2. Ibid., 157.
3. Ibid., 158.
4. Ibid., 132–33.
5. Nabhan, G., and Trimble, S. *The Geography of Childhood: Why Children Need Wild Places* (Beacon Press, 1994), 126.
6. Louv, 142–43.
7. Ibid., 158.
8. Ibid., 288.

9. Velasquez-Manoff, M. "A Cure for the Allergy Epidemic?" *New York Times* (November 10, 2013). www.nytimes.com/2013/11/10/opinion/sunday/a-cure -for-the-allergy-epidemic.html?pagewanted=2&_r=0.
10. Ibid.
11. "Screen Time and Children," a document of the National Institutes of Health (NIH). www.nlm.nih.gov/medlineplus/ency/patientinstructions/000355.htm.
12. See, e.g.: Nauert, R. "Childhood Television Watching Correlated to Later Attention Problems." *Psych Central* (September 6, 2007). http://psychcentral. com/news/2007/09/06/childhood-television-watching-correlated-to-later -attention-problems/1238.html.
13. See: Bell, D. "Protean Manifestations of Vitamin D Deficiency, Part 1: The Epidemic of Deficiency." *Southern Medical Journal* (2011). www.medscape.com/ viewarticle/742623_2.
14. "Memo to Pediatricians: Screen All Kids for Vitamin D Deficiency." Website of the Johns Hopkins Children's Hospital (February 22, 2012). www.hopkinschildrens. org/Screen-All-Kids-for-Vitamin-D-Deficiency.aspx.
15. "Vitamin D Deficiency: A Real Problem." Website of UC Berkeley Health Services. http://uhs.berkeley.edu/home/healthtopics/pdf/Vitamin%20D%20Deficiency.pdf.

Chapter 7: Health Claims Against Beef

1. American Public Health Association newsletter (January 22, 2014). http:// action.apha.org/site/MessageViewer?dlv_id=49703&em_id=45382.0.
2. Lustig, R. *Fat Chance: Beating the Odds Against Sugar, Processed Food, Obesity, and Disease* (Plume 2012), 4.
3. U.S. Centers for Disease Control and Prevention (CDC) statistics for the most recent year available (2010) show the following as the leading causes of death: (1) heart disease; (2) cancer; (3) chronic lower respiratory diseases; and (4) strokes. www.cdc.gov/nchs/fastats/lcod.htm.
4. "Coronary Artery Bypass Graft Surgery Numbers Drop 30% in 7 Years." *Medical News Today* (May 4, 2011). www.medicalnewstoday.com/articles/224100.php.
5. See "Trends in Tobacco Use." www.lung.org/finding-cures/our-research/ trend-reports/Tobacco-Trend-Report.pdf; and website of the CDC. www.cdc.gov /nchs/data/nhis/earlyrelease/earlyrelease201306_08.pdf.
6. Keirnan, B. "Grass Fed vs. Corn Fed: You Are What Your Food Eats" (July 16, 2012). www.globalaginvesting.com/news/blogdetail?contentid=1479.
7. See: "Disease Statistics," Fact Book 2012, website of the U.S. Department of Health and Human Services. www.nhlbi.nih.gov/about/factbook/chapter4.htm; and Blodget, H. "What Kills Us: The Leading Causes of Death from 1900–2010." *Business Insider* (June 2012). www.businessinsider.com /leading-causes-of-death-from-1900-2010-2012-6?op=1.
8. Parker-Pope, T. "Less Active at Work, Americans Have Packed on the Pounds." *New York Times* (May 25, 2011). http://well.blogs.nytimes.com/2011/05/25 /less-active-at-work-americans-have-packed-on-pounds/?_php=true&_ type=blogs&_r=0.
9. Reynolds, G. "The Couch Potato Goes Global." *New York Times* (July 18, 2012). http://well.blogs.nytimes.com/2012/07/18/the-couch-potato-goes-global.
10. See, for example: Brownson, R., & Boehmer, T. "Declining Rates of Physical Activity in the United States: What Are the Contributors?" *Annual Review of Public Health*, 26 (April 2005) 8, which shows that "low-activity jobs" increased from 1950 (23%) to 1970 (41%), but then leveled off. See also: Institute of Medicine, "Does the Built Environment Influence Physical Activity?" *Examining*

the Evidence, Special Report 282 (National Academies Press, 2005), which shows (p. 65) that the model T was introduced in 1908 and that by 1930 electric dish and clothes washers, dryers, and vacuums, had all been introduced; and (p. 73) that the major shift in population distribution had occurred by 1950.

11. Pollan, M. "Unhappy Meals." *New York Times Magazine* (January 28, 2007). http://michaelpollan.com/articles-archive/unhappy-meals.

12. For a recent example of the public health message advising people to cut out red meat and saturated fats, see: Kaiser Permanente newsletter (February 2014). http://partnersinhealth.kaiserpermanente.org/february-2014/national /the-surprising-truth-about-fat-natl-feb2014.

13. Keys, A. "Coronary Heart Disease in Seven Countries." *Circulation* 41, no. 1 (1970), 211.

14. Wells, H., and Buzby, J. "Dietary Assessment of Major Trends in U.S. Food Consumption, 1970–2005," a study by the Economic Research Service (ERS) of the U.S. Department of Agriculture (March 2008); Gerrior, S., et al. "Nutrient Content of the US Food Supply, 1909–2000," a study of the Center for Nutrition Policy and Promotion (CNPP), U.S. Department of Agriculture (November 2004).

15. Wells and Buzby, 2008; Gerrior, 2004.

16. Recent food consumption figures from Wells and Buzby, 2008, and Gerrior, 2004; historical consumption figures from Gerrior 2004, and the database of the Economic Research Service of the U.S. Department of Agriculture. http://ers.usda. gov/data-products/food-consumption-and-nutrient-intakes.aspx#.U6m0mqjc0Xw.

17. Enig, M. *Know Your Fats: The Complete Primer for Understanding the Nutrition of Fats, Oils, and Cholesterol* (Bethesda Press, 2000), 93.

18. ERS/CNPP/USDA data.

19. ERS/CNPP/USDA data.

20. "Per Capita Consumption," a report of the U.S. National Oceanic and Atmospheric Administration (NOAA) Office of Science and Technology, National Marine Fisheries Service. www.st.nmfs.noaa.gov/Assets/commercial/fus/fus99/per_capita99.pdf.

21. Maron, D. "Some Danish Advice on the Trans Fat Ban." *Scientific American* (November 14, 2013). http://sams.scientificamerican.com/article/ some-danish-advice-on-the.

22. "FDA Targets Trans Fat in Processed Foods," news release of the U.S. Food and Drug Administration (November 7, 2013). www.fda.gov/ForConsumers/ ConsumerUpdates/ucm372915.htm.

23. Warner, M. "A Lifelong Fight Against Trans Fat." *New York Times* (December 16, 2013). www.nytimes.com/2013/12/17/health/a-lifelong-fight-against-trans-fat .html?smid=fb-share&_r=0.

24. Sachdeva, A., et al. "Lipid Levels in Patients in Hospitalized with Coronary Artery Disease." *American Heart Journal* (January 2009). www.ahjonline.com /article/S0002-8703%2808%2900717-5/abstract.

25. Yudkin, J. *Pure, White and Deadly*, chapter 2, and 81.

26. For more about what Keys chose to exclude, see, Keith, chapter 4.

27. Kresser, C. "Red Meat: It Does a Body Good!" http://chriskresser.com/red-meat -it-does-a-body-good.

28. Ravnskov, U. "The Questionable Role of Saturated and Polyunsaturated Fatty Acids in Cardiovascular Disease." *Journal of Clinical Epidemiology* 51, no. 6 (June 1998), 443–60. www.ncbi.nlm.nih.gov/pubmed/9635993.

29. Siri-Tarino, P. W., et al. "Meta-Analysis of Prospective Cohort Studies Evaluating the Association of Saturated Fat with Cardiovascular Disease." *American Journal of Clinical Nutrition* (March 2010). www.ncbi.nlm.nih.gov/pubmed/20071648.

30. Micha, R., Wallace, S., and Mozaffarian, D. "Red and Processed Meat Consumption and Risk of Incident Coronary Heart Disease, Stroke, and Diabetes Mellitus: A Systematic Review and Meta-Analysis." *Circulation* (May 17, 2010). www.hsph.harvard.edu/news/press-releases/processed-meats-unprocessed-heart-disease-diabetes.

31. O'Connor, A. "Study Questions Fat and Heart Disease Link." *New York Times* (March 17, 2014). http://well.blogs.nytimes.com/2014/03/17/study-questions-fat-and-heart-disease-link/?_php=true&_type=blogs&nl=todays headlines&emc=edit_th_20140318&_r=0.

32. McGee, H. *On Food and Cooking: The Science and Lore of the Kitchen*, rev. ed. (Scribner, 2004), 124. Note, too, that marinating meat can substantially reduce the creation of the worrisome compounds. According to Chris Kresser, "If you do want to grill or fry your meats, you can significantly reduce the formation of all of these compounds by using an acidic marinade, which has the added bonus of tasting great! Marinating beef for one hour reduced AGE formation by over half, and marinades can cut HA formation in meat by up to 90%." See: http://chriskresser.com/does-red-meat-cause-inflammation.

33. Champeau, R. "Most Heart Attack Patients' Cholesterol Levels Did Not Indicate Cardiac Risk." *UCLA Newsroom* (January 12, 2009). http://newsroom.ucla.edu/portal/ucla/majority-of-hospitalized-heart-75668.aspx.

34. Mensink et al. "Effects of Dietary Fatty Acids and Carbohydrates on the Ratio of Serum Total to HDL Cholesterol and on Serum Lipids and Apolipoproteins: A Meta-Analysis of 60 Controlled Trials." *American Journal of Clinical Nutrition* (May 2003). www.ncbi.nlm.nih.gov/pubmed/12716665?dopt=Citation.

35. See: O'Connor, "Study Questions Fat and Heart Disease Link"; and Kresser, C. "The Diet-Heart Myth: Cholesterol and Saturated Fat Are Not the Enemy." http://chriskresser.com/the-diet-heart-myth-cholesterol-and-saturated-fat-are-not-the-enemy.

36. Ibid.

37. Mozaffarian, D. "The Optimal Diet to Prevent CVD [Cardiovascular Disease]: What Is the Role of Saturated Fat?," a plenary talk at the Health Effects of Dietary Fatty Acids Symposium, Wayne State University, Detroit, Michigan (2010). www.meandmydiabetes.com/wp-content/uploads/2010/11/Feinman-Webinar-Mozaffarian-Part-1.mp3.

38. Yudkin's clinical experiments and their results are described in detail in chapter 14 of his book *Pure, White and Deadly*.

39. Ibid.

40. Ibid., 172.

41. Ibid.

42. Smith, J. "The Man Who Tried to Warn Us Against Sugar." *Daily Telegraph* (February 13, 2014). www.calgaryherald.com/health/tried+warn+about+sugar/9503788/story.html.

43. Ibid.

44. Yudkin, 188.

45. Ibid., 113, 115.

46. Ibid., 188.

47. Smith.

48. *Nutrition Source*, website of the Harvard School of Public Health, *citing* Fung, T. T., et al. "Sweetened Beverage Consumption and Risk of Coronary Heart Disease in Women." *American Journal of Clinical Nutrition* 89 (2009), 1037. www.hsph.harvard.edu/nutritionsource/healthy-drinks/soft-drinks-and-disease.

49. *Nutrition Source*, website of the Harvard School of Public Health, *citing* de Koning, L. "Sweetened Beverage Consumption, Incident Coronary Heart Disease, and Biomarkers of Risk in Men." *Circulation* 125 (2012), 1735.

50. Malik, V. S. "Sugar-Sweetened Beverages and Risk of Metabolic Syndrome and Type 2 Diabetes: A Meta-Analysis." *Diabetes Care* 33 (2010), 2477.

51. "Soft Drinks and Disease." *Nutrition Source*, website of the Harvard School of Public Health. www.hsph.harvard.edu/nutritionsource/healthy-drinks /soft-drinks-and-disease.

52. Schmidt, L. "New Unsweetened Truths About Sugar." *Journal of the American Medical Association Internal Medicine* (April 2014). http://archinte.jamanetwork. com/article.aspx?articleid=1819571.

53. See: "Added Sugar Linked to Cardiovascular Disease." www.foodconsumer.org /newsite/2/19/sugar_heart_disease_0302140949.html.

54. Busko, M. "A Soda a Day Ups CVD Risk by 30%: NHANES Study." *Medscape* (February 4, 2014). www.medscape.com/viewarticle/820172.

55. Yang, Q., et al. "Added Sugar Intake and Cardiovascular Diseases Mortality Among US Adults." *Journal of the American Medical Association Internal Medicine* (April 2014). http://archinte.jamanetwork.com/article.aspx?articleid=1819573.

56. Hyman, M. "Fat Does Not Make You Fat." *Huffington Post* (November 26, 2013). www.huffingtonpost.com/dr-mark-hyman/fat-health_b_4343798.html.

57. Beilby, J. "Definition of Metabolic Syndrome: Report of the National Heart, Lung, and Blood Institute/American Heart Association Conference on Scientific Issues Related to Definition." *Circulation* 109 (2004), 433. www.ncbi.nlm.nih .gov/pmc/articles/PMC1880831.

58. Ibid.

59. Malhotra, A. "Sugar Is Now Enemy Number One." *Guardian* (January 11, 2014). www.theguardian.com/commentisfree/2014/jan/11/sugar-is-enemy-number -one-now.

60. Paddock, C. "Role of Saturated Fat (UK Cardiologist Questions Link)." *Medical News Today* (October 24, 2013). www.medicalnewstoday.com/articles/267834. php.

61. Beilby.

62. Musunuru, K. "Atherogenic Dyslipidemia: Cardiovascular Risk and Dietary Intervention." *Lipids* 45, no. 10 (October 2010). www.ncbi.nlm.nih.gov/pmc /articles/PMC2950930.

63. Ibid.

64. Ibid.

65. Briggs, H. "WHO: Daily Sugar Intake 'Should Be Halved.'" *BBC News* (March 5, 2014). www.bbc.com/news/health-26449497.

66. Te Morenga, L. "Dietary Sugars and Body Weight: Systematic Review and Meta- Analyses of Randomised Controlled Trials and Cohort Studies." *British Medical Journal* (2013). www.bmj.com/content/346/bmj.e7492.

67. Yang.

68. Lustig, *Fat Chance*, 7 (emphasis added).

69. Lustig, R. "The Bitter Truth," a talk at UCSF, University of California television (uploaded July 30, 2009). www.youtube.com/watch?v=dBnniua6-oM.

70. Smith.

71. "Is Sugar Toxic? *60 Minutes* Investigates." *60 Minutes* television segment (January 2013). https://www.youtube.com/watch?v=6n29ZIJ-jQA.

72. Johnson, R., et al. "Potential Role of Sugar (Fructose) in the Epidemic of Hypertension, Obesity, and the Metabolic Syndrome, Diabetes, Kidney Disease,

and Cardiovascular Disease." *American Journal of Clinical Nutrition* (October 2007). http://ajcn.nutrition.org/content/86/4/899.long#sec-3.

73. Ibid. (emphasis added).

74. Yudkin, 42.

75. On March 16, 2014, Walmart listed its price for a 5-pound bag of sugar as $2.62.

76. "World War II Rationing." *U.S. History* website. www.u-s-history.com/pages/h1674.html.

77. Sicotte, R. "The Origins and Development of the U.S. Sugar Program, 1934–1959," paper prepared for the 14th International Economic History Conference (August 2006). www.uvm.edu/~rsicotte/US%20Sugar%20Program.pdf.

78. All statistics from the United Nations Food and Agriculture Organization, from FAOSTAT at faostat.fao.org, and from the FAO report "Food Outlook: Global Market Analysis" (May 2012), 41. www.fao.org/docrep/015/al989e/al989e00.pdf.

79. Johnson.

80. Nielsen, S. "Changes in Beverage Intake Between 1977 to 2001." *American Journal of Preventive Medicine* 27 (October 2004), 205. www.ajpmonline.org/article/S0749-3797%2804%2900122-9/abstract.

81. "USC Research Finds Sodas Sweetened with More High Fructose Corn Syrup than Previously Assumed," press release of the Keck School of Medicine at USC (October 27, 2010). http://keck.usc.edu/en/About/Administrative_Offices/Office_of_Public_Relations_and_Marketing/News/Detail/archive__offices__public_relations_and_marketing__high_fructose_corn_syrup_in_sodas.

82. Ibid.

83. De Koning.

84. Also see a 2007 study published in the *New England Journal of Medicine* that connected HFCS to a variety of health problems and illnesses, including obesity, cardiovascular disease, and renal disease. www.jwatch.org/pa200712120000005/2007/12/12/high-fructose-corn-syrup-new-trans-fat.

85. Keck School of Medicine.

86. Lakhan, S. "The Emerging Role of Dietary Fructose in Obesity and Cognitive Decline." *Nutrition Journal* 12 (2013), 114. www.nutritionj.com/content/12/1/114.

87. Melville, N. "Culinary Culprits: Foods That May Harm the Brain." *Medscape* (January 30, 2014). www.medscape.com/viewarticle/819974#3.

88. Lustig, R. "The Sugar-Addiction Taboo." *The Atlantic* (January 2014). www.theatlantic.com/health/archive/2014/01/the-sugar-addiction-taboo/282699.

89. Avena, N., et al. "Evidence for Sugar Addiction: Behavioral and Neurochemical Effects of Intermittent, Excessive Sugar Intake." *Neuroscience Behavior Review* 32, no. 1 (2008). www.ncbi.nlm.nih.gov/pubmed/17617461.

90. Stice, et al. "Relative Ability of Fat and Sugar Tastes to Activate Reward, Gustatory, and Somatosensory Regions." *Journal of Clinical Nutrition* 98 (December 2013), 1377. www.ncbi.nlm.nih.gov/pubmed/24132980.

91. Lustig, *Fat Chance*, 62.

92. Sun, L. "A Mother's Food Choice Can Shape Baby's Palate, Research Shows." *Washington Post* (November 17, 2011). www.washingtonpost.com/national/health-science/a-mothers-food-choice-can-shape-babys-palate-research-shows/2011/11/17/gIQAqcYooN_story.html.

93. Fomon, S. "Infant Feeding in the 20th Century." *Journal of Nutrition* 131 (February 1, 2001). http://jn.nutrition.org/content/131/2/409S.full#FN1.

94. Lustig, *Fat Chance*, 12.

95. Institute of Medicine. *Nutrition During Lactation* (National Academies Press, 1991), 118. www.nap.edu/openbook.php?record_id=1577&page=118; see also re: the nutritional components of breast milk: www.parentingscience.com /calories-in-breast-milk.html.

96. Moskin, J. "For an All-Organic Formula, Baby, That's Sweet." *New York Times* (May 19, 2008). www.nytimes.com/2008/05/19/us/19formula.html?_r=0.

97. Terrero, N. "How Much Sugar Is in Brand-Name Baby Formula?" *NBC Latino* (February 22, 2012). http://nbclatino.com/2012/02/22/18091566837.

98. Moskin.

99. Gibbs, B. "Socioeconomic Status, Infant Feeding Practices and Early Childhood Obesity." *Pediatric Obesity* 9 (April 2014), 135. www.ncbi.nlm.nih.gov/ pubmed/23554385.

100. "AAP Reaffirms Breastfeeding Guidelines," website of the American Academy of Pediatrics (February 27, 2012). www.aap.org/en-us/about-the-aap/aap-press -room/Pages/AAP-Reaffirms-Breastfeeding-Guidelines.aspx.

101. Bartick, M, et al. "The Burden of Suboptimal Breastfeeding in the United States: A Pediatric Cost Analysis." *Pediatrics* 125 (May 2010). http://pediatrics .aappublications.org/content/125/5/e1048.full.

102. "AAP Reaffirms"; and ERS/CNPP/USDA data.

103. "Carbohydrates and Blood Sugar." *Nutrition Source*, website of the Harvard School of Public Health. www.hsph.harvard.edu/nutritionsource/carbohydrates /carbohydrates-and-blood-sugar.

104. Fillion, K. "On the Evils of Wheat." *Macleans* (September 20, 2011). www .macleans.ca/general/on-the-evils-of-wheat-why-it-is-so-addictive-and-how -shunning-it-will-make-you-skinny.

105. Jackson, N. "Your Addiction to Wheat Products Is Making You Fat and Unhealthy." *The Atlantic* (September 22, 2011). www.theatlantic.com/health/archive/2011/09/ your-addiction-to-wheat-products-is-making-you-fat-and-unhealthy/245526.

106. Mercola. "Avoid This Food to Help You Slow Aging" (February 22, 2012). http:// articles.mercola.com/sites/articles/archive/2012/02/22/how-sugar-accelerates -aging.aspx.

107. Hamblin, J. "This Is Your Brain on Gluten." *The Atlantic* (December 20, 2013). www.theatlantic.com/health/archive/2013/12/ this-is-your-brain-on-gluten/282550.

108. ERS/CNPP/USDA data.

109. Taubes, G. "What If It's All Been a Big Fat Lie?" *New York Times Magazine* (July 7, 2002). www.nytimes.com/2002/07/07/magazine/what-if-it-s-all-been-a-big -fat-lie.html?scp=1&sq=gary%20taubes%20and%20fat&st=cse.

110. "Carbohydrates." *Nutrition Source*, website of the Harvard School of Public Health (January 2014). www.hsph.harvard.edu/nutritionsource/carbohy- drate-question/?utm_source=SilverpopMailing&utm_medium= email&utm_campaign=Nutrition%20newsletter-January%202014%20 %281%29&utm_content=#low-carb.

111. Gardner, C., et al. "Comparison of the Atkins, Zone, Ornish, and LEARN Diets for Change in Weight and Related Risk Factors Among Overweight Premenopausal Women: The A to Z Weight Loss Study: A Randomized Trial." *Journal of the American Medical Association* 297 (March 7, 2007). http://jama.jamanetwork .com/article.aspx?articleid=205916.

112. "Stanford Diet Study Tips Scale in Favor of Atkins Plan, Stanford Research on Weight Loss," press release of Stanford University (March 1, 2007). http://nutrition.stanford.edu/documents/AZ_press.pdf.

113. Song, Y., et al. "A Prospective Study of Red Meat Consumption and Type 2 Diabetes in Middle-Aged and Elderly Women, the Women's Health Study." *Diabetes Care* 27 (September 2004). http://care.diabetesjournals.org /content/27/9/2108.full.

114. Ley, S. "Associations Between Red Meat Intake and Biomarkers of Inflammation and Glucose Metabolism in Women." *American Journal of Clinical Nutrition* (February 2014). http://ajcn.nutrition.org/content/early/2013/11/27 /ajcn.113.075663.abstract.

115. Kresser, C. "Does Eating Red Meat Increase the Risk of Diabetes?" chriskresser .com/does-red-meat-cause-inflammation-and-impaired-glucose-metabolism.

116. Sisson, M. "Does Eating Red Meat Increase Type 2 Diabetes Risk?" www.marksdailyapple.com/ does-eating-red-meat-increase-type-2-diabetes-risk/#ixzz2xZj5DoBh.

117. "Cleveland Clinic Researchers Discover Link Between Heart Disease and Compound Found in Red Meat, Energy Drinks," news release of Cleveland Clinic (April 7, 2013). http://my.clevelandclinic.org/media_relations /library/2013/2013-04-07-cleveland-clinic-researchers-discover-link-between -heart-disease-and-compound-found-in-red-meat-energy-drinks.aspx.

118. Pendrick, D. "New Study Links L-Carnitine in Red Meat to Heart Disease." *Harvard Health Publications* (April 17, 2013). www.health.harvard.edu/blog /new-study-links-l-carnitine-in-red-meat-to-heart-disease-201304176083.

119. Re: vegetarian diet not optimal, see: Burkert, N., et al. "Nutrition and Health—The Association Between Eating Behavior and Various Health Parameters: A Matched Sample Study." *PLOS ONE* 9, no. 2 (February 2014). www.plosone.org/article/fetchObject.action?uri=info:doi/10.1371/journal. pone.0088278&representation=PDF.

120. Kresser, C. "Red Meat and TMAO: Cause for Concern, or Another Red Herring?" http://chriskresser.com/red-meat-and-tmao-its-the-gut-not-the-meat.

121. Kresser, C. "Choline and TMAO: Eggs Still Don't Cause Heart Disease." http:// chriskresser.com/choline-and-tmao-eggs-still-dont-cause-heart-disease, *citing* De Filippo, C. "Impact of Diet in Shaping Gut Microbiota Revealed by a Comparative Study in Children from Europe and Rural Africa." *Proceedings of the National Academy of Sciences* (August 17, 2010). www.ncbi.nlm.nih.gov/pmc /articles/PMC2930426.

122. Stowkowski, L. "Red Meat and Cancer: What's the Beef?" *Medscape* (June 20, 2013). www.medscape.com/viewarticle/806573.

123. Rohrmann, S., et al. "Meat Consumption and Mortality—Results from the European Prospective Investigation into Cancer and Nutrition." *BMC Medicine* 11 (2013), 63. www.biomedcentral.com/1741-7015/11/63.

124. Alexander, D., et al. "Meta-Analysis of Prospective Studies of Red Meat Consumption and Colorectal Cancer." *European Journal of Cancer Prevention* 20, no. 4 (July 2011), 293. www.ncbi.nlm.nih.gov/pubmed/21540747; see also: Truswell, A. "Meat Consumption and Cancer of the Large Bowel." *European Journal of Clinical Nutrition* (2002). www.ncbi.nlm.nih.gov/ pubmed/11965518.

125. Missmer, S., et. al. "Meat and Dairy Food Consumption and Breast Cancer: A Pooled Analysis of Cohort Studies." *International Journal of Epidemiology* 31, no. 1 (February 2002), 78. www.ncbi.nlm.nih.gov/pubmed/11914299.

126. Kolahdooz, F. "Meat, Fish, and Ovarian Cancer Risk: Results from 2 Australian Case-Control Studies, a Systematic Review, and Meta-Analysis." *American Journal of Clinical Nutrition* 91, no. 6 (June 2010), 1752. www.ncbi.nlm.nih .gov/pubmed/20392889.

127. Rohrmann, S. "Meat and Fish Consumption and Risk of Pancreatic Cancer: Results from the European Prospective Investigation into Cancer and Nutrition." *International Journal of Cancer* 132, no. 3 (2013), 617. www.medscape.com/ medline/abstract/22610753.

128. Mozaffarian, "The Optimal Diet to Prevent CVD."

129. Gurven, M., and Kaplan, H. "Hunter Gatherer Health," a faculty paper (2007). www.anth.ucsb.edu/faculty/gurven/papers/GurvenKaplan2007pdr.pdf (internal citations omitted).

130. Ibid. (internal citations omitted).

131. McGee, 124.

132. Ibid.

133. Fallon, S., and Enig, M. "Guts and Grease: The Diet of Native Americans." Website of the Weston A. Price Foundation (January 1, 2000). www.weston aprice.org/traditional-diets/guts-and-grease.

134. Ibid. See also: Smil, V. *Should We Eat Meat?* (Wiley, 2013), especially pages 46–49, in which the author, world-renowned biologist Vaclav Smil, explains in detail how hunting preferences were made by early humans, and notes: "[T]he preference of all hunters for killing fatty animals is indisputable. This is why megaherbivores were always preferred, as greater (and often fatal) risks associated with their hunting were trumped by superior energy returns."

135. Ibid.

136. Gadsby, P., and Steel, L. "The Inuit Paradox: How Can People Who Gorge on Fat and Rarely See a Vegetable Be Healthier than We Are?" *Discover Magazine* (October 1, 2004). http://discovermagazine.com/2004/oct/inuit-paradox#.UuwgKPYhYVk.

137. Fallon and Enig.

138. Martin, D., and Goodman, A. "Health Conditions Before Columbus: A Paleopathology of Native North Americans." *Western Journal of Medicine* 176 (January 2002), 65. www.ncbi.nlm.nih.gov/pmc/articles/PMC1071659.

139. Petersen, I. "The Maasai Keep Healthy Despite a High Fat Diet." *Science Nordic* (September 11, 2012). http://sciencenordic.com/ maasai-keep-healthy-despite-high-fat-diet.

140. Masterjohn, C. "The Masai Part II: A Glimpse of the Masai Diet at the Turn of the 20th Century—A Land of Milk and Honey, Bananas from Afar." Website of the Weston A. Price Foundation (September 13, 2011). www.westonaprice.org /blogs/cmasterjohn/2011/09/13/the-masai-part-ii-a-glimpse-of-the-masai-diet -at-the-turn-of-the-20th-century-a-land-of-milk-and-honey-bananas-from-afar.

141. See also: Mann, G., et al. "Cardiovascular Disease in the Masai." *Journal of Atherosclerosis Research* 4 (July 8, 1964), 289. www.sciencedirect.com/science /article/pii/S0368131964800417.

142. Fallon, S. *Nourishing Traditions* (Newtrends Publishing, 2003), preface.

143. Diamond, J. *Guns, Germs, and Steel: The Fates of Human Societies* (W. W. Norton, 2009), 112.

Chapter 8: Beef Is Good Food

1. McGee, 123.

2. Kresser, C. "The Truth About Red Meat." http://chriskresser.com/ the-truth-about-red-meat.

3. Pobiner, B. "Evidence for Meat-Eating by Early Humans." *Nature Education Knowledge* (2013). www.nature.com/scitable/knowledge/library/evidence-for-meat-eating-by-early-humans-103874273.

4. "Iron: Dietary Supplement Fact Sheet," a National Institutes of Health fact sheet (reviewed April 8, 2014). http://ods.od.nih.gov/factsheets/Iron-HealthProfessional.

5. "Supplements: Nutrition in a Pill?" Website of the Mayo Clinic (January 19, 2013). www.mayoclinic.org/healthy-living/nutrition-and-healthy-eating/in-depth/supplements/art-20044894.

6. O'Connor, "Study Questions Fat and Heart Disease Link."

7. Nestle, M. *What to Eat* (North Point Press, 2007), 468.

8. Blythman, J., and Sykes, R. "Why Beef Is Good for You." *Guardian* (December 13, 2013). www.theguardian.com/lifeandstyle/2013/dec/16/why-beef-is-good-for-you-grass-fed-grain-fed.

9. "Beef, Nutrition Facts." *SELF Nutrition Data: Know What You Eat.* http://nutritiondata.self.com/facts/beef-products/3477/2.

10. "Nutrition: Micronutrient Deficiencies, Iron Deficiency Anaemia." Website of the World Health Organization (2014). www.who.int/nutrition/topics/ida/en.

11. Meister, A. "The Role of Beef in the American Diet," a report prepared for the American Council on Science and Health (January 2003), 10. www.meat-ims.org/wp-content/uploads/2013/07/The_Role%20_of_Beef_in_the_American_Diet.pdf.

12. "Iron: Dietary Supplement Fact Sheet."

13. "Anemia Fact Sheet." Office on Women's Health, U.S. Department of Health and Human Services (updated July 16, 2012). http://womenshealth.gov/publications/our-publications/fact-sheet/anemia.html.

14. "Iron and Iron Deficiency." U.S. Centers for Disease Control and Prevention (updated February 23, 2011). www.cdc.gov/nutrition/everyone/basics/vitamins/iron.html.

15. "Increasing Iron in Your Diet During Pregnancy." Website of the Cleveland Clinic (updated December 21, 2009). http://my.clevelandclinic.org/healthy_living/pregnancy/hic_increasing_iron_in_your_diet_during_pregnancy.aspx.

16. Ibid.

17. "Iron: Dietary Supplement Fact Sheet."

18. "Increasing Iron in Your Diet During Pregnancy."

19. Turgeon, M. *Clinical Hematology: Theory and Procedures* (LWW Publishing, 2011), 133.

20. McGee, 134.

21. "Iron: Dietary Supplement Fact Sheet."

22. Turgeon, 133.

23. "Zinc: Fact Sheet for Consumers," a National Institutes of Health fact sheet (reviewed September 20, 2011). http://ods.od.nih.gov/factsheets/Zinc-QuickFacts.

24. Meister.

25. "Zinc." Chapter 16 in *Human Vitamin and Mineral Requirements*, a joint report of the United Nations Food and Agriculture Organization and the World Health Organization (2001). www.fao.org/docrep/004/y2809e/y2809e0m.htm.

26. "Animal Protein Intake Is Associated with Higher-Level Functional Capacity in Elderly Adults: The Ohasama Study." *Journal of the American Geriatrics Society* 62, no. 3 (March 2014), 426. http://onlinelibrary.wiley.com/doi/10.1111/jgs.12690/abstract;jsessionid=B96850F343D96439D7E05E97E4DC7CEC.f03t02?systemMessage=Wiley+Online+Library+will+be+

disrupted+Saturday%2C+15+March+from+10%3A00-12%3A00+G-MT+%2806%3A00-08%3A00+EDT%29+for+essential+maintenance. Also note: Even a March 2014 study widely reported in the media as showing that middle-aged men should eat less meat actually concluded that older people who ate more meat were healthier. www.cell.com/cell-metabolism/abstract/S1550-4131%2814%2900062-X.

27. Heaney, R., and Layman, D. "Amount and Type of Protein Influences Bone Health." *American Journal of Clinical Nutrition* 87, no. 5 (May 2008). http://ajcn.nutrition.org/content/87/5/1567S.full.

28. "Animal Protein Is Good for Bones." Website of Hebrew SeniorLife, Harvard Medical School affiliate. www.hebrewseniorlife.org/research-animal-protein-is-good-for-bones.

29. Gomez-Pinilla, F. "Brain Foods: The Effects of Nutrients on Brain Function." *Nature Reviews Neuroscience* 9 (July 2008), 568. www.nature.com/nrn/journal/v9/n7/fig_tab/nrn2421_T1.html.

30. Enig, *Know Your Fats*, 57.

31. Clancy, K. *Greener Pastures: How Grass-Fed Beef and Milk Contribute to Healthy Eating*, a report of the Union of Concerned Scientists (2006), 1. www.ucsusa.org/assets/documents/food_and_agriculture/greener-pastures.pdf

32. O'Connor, A. "Low Vitamin D Levels Linked to Disease in Two Big Studies." *New York Times* (April 1, 2014). http://well.blogs.nytimes.com/2014/04/01/low-vitamin-d-levels-linked-to-disease-in-two-big-studies/?_php=true&_type=blogs&_r=0.

33. Ibid.

34. Kresser, "Red Meat: It Does a Body Good."

35. Ibid.

36. Weintraub, P. "All About B Vitamins," reprinted from *Experience Life Magazine*, website of Dr. Frank Lipman. www.drfranklipman.com/all-about-b-vitamins.

37. Ibid.

38. Ibid.

39. Ibid.; see, e.g.: Burt Berkson, PhD, MD, director of the Integrative Medicine Center of New Mexico in Las Cruces, and co-author of *User's Guide to the B-Complex Vitamins* (Basic Health Publications, 2005), who writes that most of us are woefully deficient. "Most Americans could probably benefit from taking up to three times—or more—of the RDAs for many of the B vitamins."

40. Meister, 11.

41. "Dietary Supplement Fact Sheet, Vitamin B_6," a National Institutes of Health (NIH) fact sheet (reviewed September 15, 2011). http://ods.od.nih.gov/factsheets/VitaminB6-HealthProfessional.

42. Gilsing, A., et al. "Serum Concentrations of Vitamin B_{12} and Folate in British Male Omnivores, Vegetarians and Vegans: Results from a Cross-Sectional Analysis of the EPIC-Oxford Cohort Study." *European Journal of Clinical Nutrition* 64, no. 9 (September 2010), 933. www.ncbi.nlm.nih.gov/pubmed/20648045.

43. Weintraub.

44. "Vitamin B_{12}, Fact Sheet for Consumers," a National Institutes of Health (NIH) fact sheet (reviewed June 24, 2011). http://ods.od.nih.gov/factsheets/VitaminB12-QuickFacts.

45. http://chriskresser.com/natures-most-potent-superfood.

46. See, e.g.: Hahn Niman, *Righteous Porkchop*; and Hahn Niman, N. "For Animals, Grass Each Day Keeps Doctors Away." *The Atlantic* (May 2010). www.theatlantic.com/health/archive/2010/05/for-animals-grass-each-day-keeps-doctors-away/56915.

47. McGee, 134.
48. Ibid., 135
49. Ibid., 121.
50. See, e.g.: Diez-Gonzalez, F., et al. "Grain Feeding and the Dissemination of Acid-Resistant *Escherichia coli* from Cattle." *Science* 281 (1998), 1666.
51. Clancy.
52. Ibid., 50.
53. Ibid., 25.
54. Ibid., 48.
55. Ibid.
56. Robinson, J. "Summary of Important Health Benefits of Grassfed Meats, Eggs and Dairy." www.eatwild.com/healthbenefits.htm#8 *citing* Duckett, S., et al. "Effects of Time on Feed on Beef Nutrient Composition." *Journal of Animal Science* 71, no. 8 (1993), 2079.
57. Duckett, S., et al. "Effects of Winter Stocker Growth Rate and Finishing System on: III. Tissue Proximate, Fatty Acid, Vitamin, and Cholesterol Content." *Journal of Animal Science* 87, no. 9 (June 2009), 2961. www.journalofanimalscience .org/content/87/9/2961.long.
58. Robinson, *citing British Journal of Nutrition* 105 (2011), 80.
59. Keirnan.
60. Daley, C., et al. "A Review of Fatty Acid Profiles and Antioxidant Content in Grass-Fed and Grain-Fed Beef." *Nutrition Journal* 9 (2010), 10. www.nutritionj .com/content/9/1/10.
61. Ibid.
62. Burros, M. "There's More to Like About Grassfed Beef." *New York Times* (August 30, 2006). www.nytimes.com/2006/08/30/dining/30well.html.
63. Curry, L. *Pure Beef: An Essential Guide to Artisan Meat with Recipes for Every Cut* (Running Press, 2012), 58.
64. McGee, 138.
65. "Tackling Climate Change," FAO.
66. Pimentel and Pimentel, 129.
67. Snell, M. "Medieval Food Preservation." http://historymedren.about.com/od /foodandfamine/a/food_preservation.htm.
68. McGee, 145.
69. Danforth, A. *Butchering Beef: The Comprehensive Photographic Guide to Humane Slaughtering and Butchering* (Storey Publishing, 2014), 93.

A Critique: What's the Matter with Beef?

1. Survey conducted by Mintel, MeatPoultry.com (January 24, 2014). www.meat poultry.com/articles.
2. "Cargill's View on Zilmax Being Pulled from the Market." Cargill News Center. www .cargill.com/news/cargill-view-on-zilmax-being-pulled-from-the-market/index.jsp.
3. "Concerns Raised About Using Beta Agonists in Beef Cattle." *Science Daily* (March 12, 2014). www.sciencedaily.com/releases/2014/03/140312181913.htm.
4. Grandin, T. "The Effect of Economics on the Welfare of Cattle, Pigs, Sheep, and Poultry." Website of Temple Grandin (April 2013). www.grandin.com/welfare /economic.effects.welfare.html.
5. Charles, D. "Inside the Beef Industry's Battle Over Growth-Promotion Drugs." Website of National Public Radio (August 21, 2013). www.npr.org/blogs/ thesalt/2013/08/21/214202886/inside-the-beef-industrys-battle-over -growth-promotion-drugs.

6. See: Mellon, M. *Hogging It: Estimates of Antimicrobial Abuse in Livestock*, a report of the Union of Concerned Scientists (2001).
7. See, e.g.: "It is well established that predators play a vital role in maintaining structure and stability of communities and that removal of predators can have a variety of cascading, indirect effects" (Terborgh et al., 2001; Duffy, 2003). Krausman.
8. See: Hahn Niman, *Righteous Porkchop*, for a more complete discussion of this issue.
9. "Standards." Website of Animal Welfare Approved, a program of the Animal Welfare Institute. http://animalwelfareapproved.org/standards /beef-cattle-2014/#130-transport.

Final Analysis: Why Eat Animals?

1. Nestle, M. *Food Politics* (University of California Press, 2002), 13.
2. Kimbrell, A. *The Fatal Harvest Reader: The Tragedy of Industrial Agriculture* (Island Press, 2002), 7.
3. McLaughlin, M. *World Food Security* (Center of Concern, 2002).
4. Ibid.
5. Eisler, M., et al. "Agriculture: Steps to Sustainable Livestock." *Nature* 507 (March 2014). www.nature.com/news/ agriculture-steps-to-sustainable-livestock-1.14796.
6. "Tackling Climate Change," FAO, 1.
7. Pathak, K. "Livestock Development: How It Contributes to Smallholder Farms." *Food Tank* (March 6, 2014). http://foodtank.com/news/2014/03/ livestock-development-how-it-contributes-to-smallholder-farmers.
8. Fairlie, 109, *quoting* Dando, *The Geography of Famine* (V. H. Winston and Sons, 1980).
9. McGee, 122.
10. Pathak.
11. Eisler. See, for example: "[O]xen consume mostly forage, which is unsuitable for human consumption. ... In Guatemala, the use of about 310h of ox power reduces the human labor input almost by half." Pimentel and Pimentel, 103.
12. Tudge, C. *Feeding People Is Easy* (Pari Publishing, 2007), 67.
13. www.rural21.com/english/news/detail/article/ livestock-futures-conference-about-powerlessness-and-hope-0000466.
14. Pimentel and Pimentel, 65.
15. Ibid.
16. Eisler.
17. Ibid.
18. Pimentel and Pimentel, 73.
19. Ibid., 69.
20. Montgomery, 35.
21. Pimentel and Pimentel, 62.
22. Troeh, F., et al. *Soil and Water Conservation: Productivity and Environmental Protection*, 3rd ed. (Prentice-Hall, 1999), 337.
23. Rinehart, L. "Ruminant Nutrition for Graziers." National Sustainable Agriculture Information Service, (2008). https://attra.ncat.org/attra-pub/viewhtml.php?id=201.
24. Tudge, 67.
25. *California Ecosystems*, 2.
26. Ibid.
27. Pimentel and Pimentel, 58, 59.
28. Ibid., 74.
29. www.beefboard.org/news/files/factsheets/The-Environment-And-Cattle -Production.pdf.

30. *California Ecosystems*, 2.
31. Ibid., 4.
32. Ibid., 14.
33. Fairlie, 30.
34. Ibid., 36.
35. Hahn Niman, N. "Can Meat Eaters Also Be Environmentalists?" *The Atlantic* (June 2, 2010). www.theatlantic.com/health/archive/2010/06 /can-meat-eaters-also-be-environmentalists/57532.
36. Hahn Niman, "Animals Are Essential."
37. Hahn Niman, N. "Dogs Aren't Dinner: The Flaws in an Argument for Veganism." *The Atlantic* (November 4, 2010). www.theatlantic.com/health/archive/2010/11/ dogs-arent-dinner-the-flaws-in-an-argument-for-veganism/66095.
38. See, e.g.: Keith; and Dunn, R. "Human Ancestors Were Nearly All Vegetarians." *Scientific American* (July 23, 2012). http://blogs.scientificamerican.com/ guest-blog/2012/07/23/human-ancestors-were-nearly-all-vegetarians.
39. Fearnley-Whittingstall, H. *The River Cottage Meat Book* (Ten Speed Press, 2007), 12.
40. Ibid., 16.
41. Howard, A. "The Animal as Our Farming Partner." *Organic Gardening* 2, no. 3 (September 1947). http://journeytoforever.org/farm_library/howard_animal .html.
42. Fairlie, 270.
43. Ibid., 273.
44. Ibid., 273.
45. Troeh, 349.
46. Pimentel and Pimentel, 364.

INDEX

ABOUT THE AUTHOR

MITCH TOBIAS

Nicolette Hahn Niman previously served as senior attorney for the Waterkeeper Alliance, running its campaign to reform the industrialized production of livestock and poultry. In recent years she has gained a national reputation as an advocate for sustainable food production and improved farm animal welfare. She is the author of *Righteous Porkchop* (HarperCollins, 2009) and has written for numerous publications, including *The New York Times*, the *Los Angeles Times*, *The Huffington Post*, and *The Atlantic* online. She lives on a ranch in Bolinas, California, with her husband, Bill Niman, and their two sons, Miles and Nicholas.

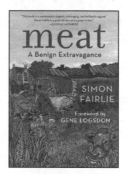